Domain-Specific Languages in R

Advanced Statistical Programming

Thomas Mailund

Apress®

Domain-Specific Languages in R: Advanced Statistical Programming

Thomas Mailund
Aarhus N, Staden København, Denmark

ISBN-13 (pbk): 978-1-4842-3587-4   ISBN-13 (electronic): 978-1-4842-3588-1
https://doi.org/10.1007/978-1-4842-3588-1

Library of Congress Control Number: 2018947391

Managing Director, Apress Media LLC: Welmoed Spahr
Acquisitions Editor: Steve Anglin
Development Editor: Matthew Moodie
Coordinating Editor: Mark Powers

Cover designed by eStudioCalamar

Cover image designed by Freepik (www.freepik.com)

Distributed to the book trade worldwide by Springer Science+Business Media New York, 233 Spring Street, 6th Floor, New York, NY 10013. Phone 1-800-SPRINGER, fax (201) 348-4505, e-mail orders-ny@springer-sbm.com, or visit www.springeronline.com. Apress Media, LLC is a California LLC and the sole member (owner) is Springer Science + Business Media Finance Inc (SSBM Finance Inc). SSBM Finance Inc is a Delaware corporation.

For information on translations, please e-mail editorial@apress.com; for reprint, paperback, or audio rights, please e-mail bookpermissions@springernature.com.

Apress titles may be purchased in bulk for academic, corporate, or promotional use. eBook versions and licenses are also available for most titles. For more information, reference our Print and eBook Bulk Sales web page at www.apress.com/bulk-sales.

Any source code or other supplementary material referenced by the author in this book is available to readers on GitHub via the book's product page, located at www.apress.com/9781484235874. For more detailed information, please visit www.apress.com/source-code.

Printed on acid-free paper

Table of Contents

About the Author

Thomas Mailund is an associate professor in bioinformatics at Aarhus University, Denmark. He has a background in math and computer science. For the past decade, his main focus has been on genetics and evolutionary studies, particularly comparative genomics, speciation, and gene flow between emerging species. He has published *Beginning Data Science in R*, *Functional Programming in R*, and *Metaprogramming in R*, all from Apress, as well as other books.

About the Technical Reviewer

 **Colin Fay** works for ThinkR, a French agency focused on everything R-related.

During the day, he helps companies take full advantage of the power of R by providing training (from beginner to expert), tools (packages, shiny apps...), and infrastructure. His main areas of expertise are software engineering, analytics, and data visualization.

During the night, Colin is a hyperactive open source developer and an open data advocate. You can find a lot of his work on his GitHub account (https://github.com/ColinFay).

He is also active in the data science community in France, especially in his home town Rennes, where he founded the data-blogging web site Data-Bzh.fr, co-founded the Breizh Data Club association, and organizes the Breizh Data Club Meetups.

You can learn more about Colin via his web site at https://colinfay.me, and you can find him on Twitter at https://twitter.com/_ColinFay.

To learn more about ThinkR, please visit www.thinkr.fr/, https://github.com/ThinkR-open, and https://twitter.com/Thinkr_FR.

CHAPTER 1

Introduction

This book introduces embedded domain-specific languages in R. The term *domain-specific languages,* or DSL, refers to programming languages specialized for a particular purpose, as opposed to general-purpose programming languages. Domain-specific languages ideally give you a precise way of specifying tasks you want to do and goals you want to achieve, within a specific context. Regular expressions are one example of a domain-specific language, where you have a specialized notation to express patterns of text. You can use this domain-specific language to define text strings to search for or specify rules to modify text. Regular expressions are often considered very hard to read, but they do provide a useful language for describing text patterns. Another example of a domain-specific language is SQL—a language specialized for extracting from and modifying a relational database. With SQL, you have an expressive domain-specific language in which you can specify rules as to which data points in a database you want to access or modify.

Who This Book Is For

This book is aimed at experienced R programmers. Some of the concepts we cover in this book are advanced, so at the very least you should be familiar with functional and object-oriented programming in R (although the next chapter will review the object-oriented programming features

© Thomas Mailund 2018
T. Mailund, *Domain-Specific Languages in R,*
https://doi.org/10.1007/978-1-4842-3588-1_1

we will use). It will be helpful to have some experience with meta-programming when it comes to evaluating expressions in contexts that interact with the surrounding R code. However, Chapter 7 gives a crash course in the techniques we will use in this book, so you should be able to pick it up with a little effort from there.

Domain-Specific Languages

With domain-specific languages we often distinguish between "external" and "embedded" languages. Regular expressions and SQL are typically specified as strings when you use them in a program, and these strings must be parsed and interpreted when your program runs. In a sense, they are languages separated from the programming language you use them *in*. They need to be compiled separately and then called by the main programming language. They are therefore considered "external" languages. In contrast, *embedded* domain-specific languages provide domain-specific languages expressed in the general-purpose language in which they are used. In R, the grammar of graphics implemented in ggplot2 or the data transformation operations implemented in dplyr provides small languages—domain-specific languages—that you can use from within R, and you write the programs for these languages in R as well.

Embedded DSLs extend the programming language in which you are working. They provide more than what you usually find in a framework in the form of functions and classes as they offer a high level of flexibility in what you can do with them. They are programming languages, after all, and you can express complex ideas and tasks in them. They provide a language for expressing thoughts in a specific domain, so they do not give you a general programming language as the language you use them from, but they do extend that surrounding language with new expressive power. However, being embedded in the general-purpose language means that they will follow the rules you are familiar with there—or mostly, at least, since in languages such as R it is possible to modify the rules a bit using so-called *non-standard evaluation*. You can expect the syntax of the embedded DSL to follow the rules of the general-purpose language. The semantics will be determined by the DSL, but when you write programs in the DSL, the syntax is already familiar to you. If you implement a DSL yourself, embedding it in a general-purpose language lets you reuse the parsing and evaluation done by the general-purpose language so that you can focus on designing the domain-specific language.

Implementing embedded domain-specific languages often involves *meta-programming*; that is, it consists of treating the program you are writing as data that you can manipulate from within the language itself. This might sound more complicated than it is, but quite often, it is reasonably straightforward to achieve. Using classes and operator overloading, we can use R's parser to parse embedded languages by simply designing the language such that evaluating expressions automatically parse them. This leaves us with data structures we, ourselves, have defined, without us having to parse anything, and we can rewrite the results of such parsed expressions in various ways before we evaluate them to run the embedded program. Evaluating expressions can be relatively straightforward or involve a deeper analysis of the parsed expressions.

To get a feel for what we can do with embedded domain-specific languages, let's consider a simple DSL: matrix multiplication (an example we cover in more detail in Chapter 2). You might not think of matrix expressions as much of a programming language, but the arithmetic notation is highly efficient for expressing ideas in a limited domain. Just imagine having to do mathematics without this notation. Of course, R already supports this language—it has infix operators and the semantics we associate with arithmetic expressions when we write them in an R program. However, since matrix multiplication is a well-known task, it serves as an excellent example to illustrate some of the things we can do if we extend R with other smaller programming languages.

R already supports arithmetic with matrices, and if you use the operator %*%, you can do matrix multiplication (if you use *, you will do component-wise multiplication instead). Multiplications are done one at a time, so if you have a series of them, such as this:

```
A %*% B %*% C %*% D
```

then the product will be computed from left to right, like this:

```
((A %*% B) %*% C) %*% D
```

For each multiplication, you produce a new matrix that will be used in the next multiplication.

Now, matrix multiplication is associative, so you should be able to set the parentheses in any way, as long as you respect the left-to-right order of the matrices (matrix multiplication is not commutative, after all), and you will get the same result. The running time, however, will not be the same. We can do a small experiment to see this using the microbenchmark package.

```
A <- matrix(1, nrow = 400, ncol = 300)
B <- matrix(1, nrow = 300, ncol = 30)
C <- matrix(1, nrow = 30, ncol = 500)
D <- matrix(1, nrow = 500, ncol = 400)

library(microbenchmark)
res <- microbenchmark(A %*% B %*% C %*% D,
                      ((A %*% B) %*% C) %*% D,
                      (A %*% (B %*% C)) %*% D,
                      (A %*% B) %*% (C %*% D),
                      A %*% (B %*% (C %*% D)),
                      A %*% ((B %*% C) %*% D))

options(microbenchmark.unit="relative")
print(res, signif = 3, order = "mean")
```

```
## Unit: relative
##                         expr  min   lq mean median
## (A %*% B) %*% (C %*% D) 1.00 1.00 1.00   1.00
## A %*% (B %*% (C %*% D)) 3.92 3.87 3.49   3.84
##       A %*% B %*% C %*% D 6.13 6.06 5.42   6.03
## ((A %*% B) %*% C) %*% D 6.12 6.05 5.51   6.04
## A %*% ((B %*% C) %*% D) 7.71 7.62 6.75   7.57
## (A %*% (B %*% C)) %*% D 9.88 9.76 8.73   9.69
##   uq  max neval
## 1.00 1.00   100
## 3.62 1.41   100
## 5.57 2.06   100
## 5.61 2.35   100
## 7.00 2.30   100
## 8.99 3.71   100
```

Here, I've computed the matrix product in the five different possible ways. There are six expressions, but the first two will compute the matrix multiplication in the same order. With microbenchmark we compute each expression 100 times and collect the time each evaluation takes. We collect the time it takes to compute each expression, and here I have displayed the running time relative to the fastest expression, sorted by the mean evaluation time.

On average, there is almost a factor of ten between the fastest and the slowest evaluation (for the slowest evaluations in the two cases the difference is a factor of two, which is still a substantial relative difference). There is something to be gained by setting parentheses optimally if we multiply together several large matrices. The dimensions of matrices are not necessarily known before runtime, however, so ideally we want to set the parentheses when we evaluate expressions in an optimal way.

The approach we take in Chapter 2 is to delay the evaluation of matrix multiplication and instead build a data structure for matrix expressions, one we can evaluate later when we have the entire matrix multiplication expression constructed. It is a simple DSL, but it contains all the components we typically need in one: we need code for parsing an expression and creating a representation of it, we need to do some manipulation of expressions, and then we need to evaluate them.

For parsing expressions, we need to capture matrices and multiplications. We wrap matrices in a class to make them objects of our language, rather than plain R data.

```r
m <- function(data) {
  structure(data,
            nrow = nrow(data),
            ncol = ncol(data),
            class = c("matrix_expr", class(data)))
}
```

The class matrix_expr is one we will use to overload the multiplication operator, and we want all elements of this class to know their dimensions, so when we wrap the data, we save the number of rows, nrow, and the number of columns, ncol.

We also want to capture matrix multiplications, where we do not evaluate the multiplication but simply save references to the matrices that we want to multiply together.

```r
matrix_mult <- function(A, B) {
  structure(list(left = A, right = B),
            nrow = nrow(A),
            ncol = ncol(B),
            class = c("matrix_mult", "matrix_expr"))
}
```

Of course, we do not want to write expressions as follows:

```
matrix_mult(matrix_mult(m(A), m(B), matrix_mult(m(C), m(D))))
```

so we overload the * operator:

```
`*.matrix_expr` <- function(A, B) {
  matrix_mult(A, B)
}
```

Now, this expression:

```
m(A) * m(B) * m(C) * m(D)
```

is our new syntax for the matrix multiplication, shown here:

```
A %*% B %*% C %*% D
```

except that the former expression only constructs a data structure representing the expression. It does not evaluate it.

When we need to evaluate a matrix multiplication, we want to analyze the delayed evaluation and rearrange the multiplication to get the optimal performance. In Chapter 2 we will implement the functions rearrange_ matrix_mult and eval_matrix_mult that do this. Here, we just define a function, v, for evaluating a matrix multiplication:

```
v <- function(expr)
    eval_matrix_expr(rearrange_matrix_expr(expr))
```

We can compare this automatic parentheses setting procedure with the default evaluation and the optimal evaluation order we saw earlier.

```
res <- microbenchmark(A %*% B %*% C %*% D,
                      (A %*% B) %*% (C %*% D),
                      v(m(A) * m(B) * m(C) * m(D)))

options(microbenchmark.unit="relative")
print(res, signif = 3, order = "mean")
```

7

```
## Unit: relative
##                               expr  min    lq mean
##        (A %*% B) %*% (C %*% D) 1.00 1.00 1.00
## v(m(A) * m(B) * m(C) * m(D)) 1.13 1.19 1.37
##           A %*% B %*% C %*% D 6.13 6.09 5.65
## median   uq  max neval
##   1.00 1.00 1.00   100
##   1.23 1.26 1.19   100
##   6.08 5.99 2.19   100
```

The automatic solution is only slightly slower than the optimal solution and about a factor of six better than the default evaluation.

CHAPTER 2

Matrix Expressions

In the next chapter we discuss computer languages in a more theoretical way, but here we will consider a concrete case—the matrix expressions mentioned in Chapter 1. This example is a relatively simple domain-specific language, but parsing matrix expressions, optimizing them, and then evaluating them are all the phases we usually have to implement in any DSL, and the implementation will also have examples of most of the techniques we will cover in more detail later. The example will use some tricks that I will not explain until later in the book, so some aspects might not be evident at this point, but the broader strokes should be and will ideally serve as a sneak peek of what follows in future chapters.

Our goal for writing a language for matrix expressions is to improve upon the default performance of the built-in matrix expressions. We achieve this by taking a more global view of expressions than R does—R will handle each operator one at a time from left to right, but we will analyze expressions and rearrange them to improve performance. These are the steps we must take to do this:

1. Parse expressions into data that we can manipulate.

2. Rearrange the expressions into more efficient expressions.

3. Provide a way to evaluate the expressions.

© Thomas Mailund 2018
T. Mailund, *Domain-Specific Languages in R*,
https://doi.org/10.1007/978-1-4842-3588-1_2

In this chapter, we use the following library:

```
library(microbenchmark)
```

Parsing Expressions

To keep things simple, we will only consider matrix multiplication and matrix addition. We do not include scalar multiplication or inverting or transposing matrices or any other functionality. Adding more components of the expression language in the example will follow the same ideas as we need for multiplication and addition. It will not teach us anything new regarding embedding DSLs in R. When you understand the example, you will be able to do this easily yourself.

With these restrictions, we can say that a matrix expression is either just a matrix, the product of two matrix expressions, or the sum of two matrix expressions. We can represent this as a class hierarchy with one (abstract) superclass representing expressions and three (concrete) subclasses for actual data, products, and sums. If you are not familiar with object-oriented programming in R, we will have a short guide to everything you need to know in Chapter 4. Constructors for creating objects of the three concrete classes can look like these:

```
m <- function(data) {
  structure(list(data = data),
            nrow = nrow(data),
            ncol = ncol(data),
            def_expr = deparse(substitute(data)),
            class = c("matrix_data", "matrix_expr"))
}
matrix_mult <- function(A, B) {
```

```
  structure(list(left = A, right = B),
              nrow = nrow(A),
              ncol = ncol(B),
              class = c("matrix_mult", "matrix_expr"))
}
matrix_sum <- function(A, B) {
  structure(list(left = A, right = B),
              nrow = nrow(A),
              ncol = ncol(B),
              class = c("matrix_sum", "matrix_expr"))
}
```

We just wrap the parameters of the constructors in a list and set the appropriate class attributes, and we store the number of rows and number of columns because we will need them when optimizing matrix multiplication, as we saw in Chapter 1.

The only purpose of the def_expr attribute we set in the m function is pretty printing. It makes the output of the expressions we manipulate in the following pages easier to follow. Strictly speaking, we do not *need* any pretty printing for manipulating expressions, but it does make debugging easier, so I tend always to write some code for that. For the matrix expressions, we can use the following code:

```
toString.matrix_data <- function(x, ...) {
  paste0("[", attr(x, "def_expr"), "]")
}
toString.matrix_mult <- function(x, ...) {
  paste0("(", toString(x$left), " * ", toString(x$right), ")")
}
toString.matrix_sum <- function(x, ...) {
  paste0("(", toString(x$left), " + ", toString(x$right), ")")
}
print.matrix_expr <- function(x, ...) {
```

```
  cat(toString(x), "\n")
}
```

Using the constructors and the pretty-printing code, we can try to construct a small expression.

```
A <- matrix(1, nrow = 10, ncol = 20)
B <- matrix(1, nrow = 20, ncol = 10)
C <- matrix(1, nrow = 10, ncol = 10)

matrix_sum(matrix_mult(m(A), m(B)), m(C))

## (([A] * [B]) + [C])
```

There is nothing in what we have done so far that qualifies as providing a *language* as such. We have just implemented a few constructor functions. However, if we overload the multiplication and addition operators for matrix expressions, we get something that starts to resemble a language at least.

```
`*.matrix_expr` <- function(A, B) {
  stopifnot(ncol(A) == nrow(B))
  matrix_mult(A, B)
}
`+.matrix_expr` <- function(A, B) {
  stopifnot(dim(A) == dim(B))
  matrix_sum(A, B)
}
```

With these, we can write the same expression more familiarly.

```
m(A) * m(B) + m(C)

## (([A] * [B]) + [C])
```

I have put some assertions—the calls to stopifnot()—into the code for the operators to make sure that the dimensions of the matrices involved in operators are valid. We *could* also have placed these in the constructor functions, but later, we will manipulate expressions where we know that the dimensions are valid, so we do not need to check them there. We do not expect a user to call the constructors directly but use the operators, so this is the natural place to put the checks.

We use the dim function for the sanity check in the addition operator, so we need a version of this that works on matrix expressions. It could look like this:

```
dim.matrix_expr <- function(x) {
  c(attr(x, "nrow"), attr(x, "ncol"))
}
```

You might be wondering why we need the m function. After all, it does not contribute anything to expressions instead of just wrapping matrices. Could we just use the matrices directly? The answer is no, and it has to do with how we use operator overloading. For * and + to be the matrix expression versions, we need the first arguments given to them to be a matrix expression. If we wrote simply this:

```
A * B + C
```

```
## Error in A * B: non-conformable arrays
```

we would be invoking the operators for R's matrix class instead. And since * is not matrix multiplication (for that you need to use %*% because the * operator is component-wise multiplication), you get an error.

We need a way of bootstrapping us from R's matrices to the matrices in our expression language. That is what we use m for.

Meta-Programming Parsing

Using an explicit function such as m to bootstrap us into the matrix expression language is the simplest way to use R's own parser for our benefits, but it is not the only way. In R, we can manipulate expressions as if they were data, a feature known as *meta-programming* and something we return to in Chapter 5. For now, it suffices to know that an expression can be explored recursively. We can use the predicate is.name to check whether the expression refers to a variable, and we can use the predicate is.call to check whether it is a function call—and all operators are function calls. So, given an expression that does not use the m function and thus does not enter our DSL, we can transform it into one that goes like this:

```
build_matrix_expr <- function(expr) {
  if (is.name(expr)) {
    return(substitute(m(name), list(name = expr)))
  }

  if (is.call(expr)) {
    if (expr[[1]] == as.name("("))
      return(build_matrix_expr(expr[[2]]))
    if (expr[[1]] == as.name("*") ||
        expr[[1]] == as.name("%*%")) {
      return(call('*',
                  build_matrix_expr(expr[[2]]),
                  build_matrix_expr(expr[[3]])))
    }
    if (expr[[1]] == as.name("+")) {
      return(call('+',
                  build_matrix_expr(expr[[2]]),
                  build_matrix_expr(expr[[3]])))
    }
  }
}
```

```
  stop(paste("Parse error for", expr))
}
```

In this implementation, we consider both * and %*% matrix multiplication so that we would consider an R expression that uses matrix multiplication as such. Notice also that we consider calls that are parentheses. Parentheses are also function calls in R, and if we want to allow our language to use parentheses, we have to deal with them—like here, where we just continue the recursion. We did not have to worry about that when we explicitly wrote expressions using m and operator overloading because there R already took care of giving parentheses the right semantics.

For this function to work, it needs a so-called quoted expression. If we write a raw expression in R, then R will try to evaluate it before we can manipulate it. We will get an error before we even get to rewrite the expression.

```
build_matrix_expr(A * B)
```

```
## Error in A * B: non-conformable arrays
```

To avoid this, we need to quote the expression.

```
build_matrix_expr(quote(A * B))
```

```
## m(A) * m(B)
```

We can avoid having to explicitly quote expressions every time we call the function by wrapping it in another function that does this for us. If we call the function substitute on a function parameter, we get the expression it contains so that we can write a function like this:

```
parse_matrix_expr <- function(expr) {
  expr <- substitute(expr)
  build_matrix_expr(expr)
}
```

15

Now, we do not need to quote expressions to do the rewriting.

parse_matrix_expr(A * B)

m(A) * m(B)

This isn't a perfect solution, and there are some pitfalls, among which is that you cannot use this function from other functions directly. The substitute function can be difficult to work with. The further problem is that we are creating a new expression, but it's an R expression and not the data structure we want in our matrix expression language. You can think of the R expression as a literate piece of code; it is not yet evaluated to become the result we want. For that, we need the eval function, and we need to evaluate the expression in the right context. Working with expressions, especially evaluating expressions in different environments, is among the more advanced aspects of R programming, so if it looks complicated right now, do not despair. We cover it in detail in Chapter 7. For now, we will just use this function:

```
parse_matrix_expr <- function(expr) {
  expr <- substitute(expr)
  modified_expr <- build_matrix_expr(expr)
  eval(modified_expr, parent.frame())
}
```

It gets the (quoted) expression, builds the corresponding matrix expression, and then evaluates that expression in the "parent frame," which is the environment where we call the function. With this function, we can get a data structure in our matrix language from an otherwise ordinary R expression.

parse_matrix_expr(A * B)

([A] * [B])

The approach we take here involves translating one R expression into another to use our m function to move us from R to matrix expressions. This involves parsing the expression twice, once when we transform it and again when we ask R to evaluate the result. The approach is also less expressive than using the m function directly. We can call m with any expression that generates a matrix, but in the expression transformation, we only allow identifiers.

As an alternative, we can build the matrix expression directly using our constructor functions. We will use matrix_mult and matrix_sum when we have a call that is *, %*%, or +, and otherwise, we will call m. This way, any expression we do not recognize as multiplication or addition will be interpreted as a value we should consider a matrix. This approach, however, adds one complication. When we call function m, we need to call it with a value, but what we have when traversing the expression is *quoted* expressions. We need to evaluate such expressions, and we need to do so in the right environment. We will need to pass an environment along with the traversal for this to work.

```
build_matrix_expr <- function(expr, env) {
  if (is.call(expr)) {
    if (expr[[1]] == as.name("("))
      return(build_matrix_expr(expr[[2]], env))
    if (expr[[1]] == as.name("*") || expr[[1]] == as.name("%*%"))
      return(matrix_mult(build_matrix_expr(expr[[2]], env),
                         build_matrix_expr(expr[[3]], env)))
    if (expr[[1]] == as.name("+"))
      return(matrix_sum(build_matrix_expr(expr[[2]], env),
                        build_matrix_expr(expr[[3]], env)))
  }
  data_matrix <- m(eval(expr, env))
  attr(data_matrix, "def_expr") <- deparse(expr)
  data_matrix
}
```

Most of this function should be self-explanatory, except for where we explicitly set the def_expr attribute of a data matrix. This is the attribute to be used for pretty printing, and when we call the m function, it is set to the literate expression we called m with. This would be eval(expr, env) for all matrices we create with this function. To avoid that, we explicitly set it to the expression we use in the evaluation.

Once again, we can wrap the function in another that gets us the quoted expression and provide the environment in which we should evaluate expressions.

```
parse_matrix_expr <- function(expr) {
  expr <- substitute(expr)
  build_matrix_expr(expr, parent.frame())
}

parse_matrix_expr(A * B + matrix(1, nrow = 10, ncol = 10))

## (([A] * [B]) + [matrix(1, nrow = 10, ncol = 10)])
```

There is more to know about manipulating expressions, especially about how they are evaluated, but we will return to that in later chapters.

Expression Manipulation

Our goal for writing this matrix DSL is to optimize evaluation of these matrix expressions. There are several optimizations we can consider, but R's matrix implementation is reasonably efficient already. It is hard to beat if we try to replace any computations by our own implementations—at least as long as we implement our alternatives in R. Therefore, it makes sense to focus on the arithmetic rewriting of expressions.

We can rewrite expressions recursively and use a generic function with specializations for the three concrete classes we have. A template (that does not do anything yet) would look like this:

```
rearrange_matrix_expr <- function(expr) {
  UseMethod("rearrange_matrix_expr")
}
rearrange_matrix_expr.matrix_data <- function(expr) {
  expr
}
rearrange_matrix_expr.matrix_mult <- function(expr) {
  matrix_mult(rearrange_matrix_expr(expr$left),
              rearrange_matrix_expr(expr$right))
}
rearrange_matrix_expr.matrix_sum <- function(expr) {
  matrix_sum(rearrange_matrix_expr(expr$left),
             rearrange_matrix_expr(expr$right))
}
```

These functions traverse a matrix expression and return the same expression structure. We can modify the functions based on patterns of expressions, however, to start rearranging.

We can make some reasonable guesses at how many operations are needed to evaluate an expression from these two rules: 1) multiplying an $n \times k$ matrix to a $k \times m$ matrix involves $n \times k \times m$ operations, and 2) adding two $n \times m$ matrices together involves $n \times m$ operations. If we can do any rewriting of an expression that reduces the number of operations we have to do, then we are improving the expression.

There are some obvious patterns we could try to match and rewrite. For instance, we should always prefer $(A + B)C$ over $AC + BC$. However, we can probably expect that the programmer writing an expression already knows this, so we will likely get little to gain from such obvious rewrites. Where we might get some performance improvements is when expressions consist of several matrices multiplied together. There, the order of multiplications matters for the number of operations we have to perform, and the optimal order depends on the dimensions of the matrices; we cannot merely look at the arithmetic expression and see the obvious way of setting parentheses to get the best performance.

Optimizing Multiplication

Before we start rewriting multiplication expressions, though, we should figure out how to find the optimal order of multiplication. Let's assume that we have this matrix multiplication: $A_1 \times A_2 \times ... \times A_n$. We need to set parentheses somewhere, say $(A_1 \times A_2 \times ...A_i) \times (A_{i+1}...\times A_n)$, to select the *last* matrix multiplication. If we first multiply together, in some order, the first i and the last $n - i$ matrices, the last multiplication we have to do is the product of those two. If the dimensions of $(A_1 \times ...A_i)$ are $n \times k$ and the dimensions of $(A_{i+1}...\times A_n)$ are $k \times m$, then this approach will involve $n \times k \times m$ operations plus how long it takes to produce the two matrices. Assuming that the best possible way of multiplying the first i matrices involves $N_{1,i}$ operations and assuming the best possible way of multiplying the last $n - i$ matrices together involves $N_{i+1,n}$ operations, then the best possible solution that involves setting the parentheses where we just did involves $N_{1,i} + N_{i+1,n} + n \times k \times m$ operations. Obviously, to get the best performance, we must pick the best i for setting the parentheses at the top level, so we must minimize this expression for i. Recursively, we can then solve for the sequences 1 to i and $i + 1$ to n to get the best performance.

Put another way, the minimum number of operations we need to multiply matrices $A_i, A_{i+1}, ..., A_j$ can be computed recursively as $N_{i,j} = 0$ when $i = j$ and

$$N_{i,j} = \min_k \left\{ N_{i,k} + N_{k+1,j} + \mathrm{nrow}(A_i) \times \mathrm{ncol}(A_k) \times \mathrm{ncol}(A_j) \right\}$$

otherwise. Actually computing this recursively would involve recomputing the same values many times, but using dynamic programming we can compute the $N_{i,j}$ table efficiently, and from that table we can backtrack and find the optimal way of setting parentheses as well.

In the following implementation, we assume that we have such a list of matrices as input. We then collect their dimensions in a table, dims, for easy access. Then, we simply create a table to represent the $N_{i,j}$ values and fill it using the previous equation. Once we have filled the table, we backtrack through it to work out the optimal way of multiplying together the matrices from 1 to n, given the dimensions, table, and matrices.

```
arrange_optimal_matrix_mult <- function(matrices) {
  n <- length(matrices)
  dims <- matrix(0, nrow = n, ncol = 2)
  for (i in seq_along(matrices)) {
    dims[i,] <- dim(matrices[[i]])
  }

  N <- matrix(0, nrow = n, ncol = n)
  for (len in 2:n) {
    for (i in 1:(n - len + 1)) {
      j <- i + len - 1
      k <- i:(j - 1)
      N[i,j] <- min(dims[i,1]*dims[k,2]*dims[j,2] + N[i,k] +
      N[k + 1,j])
    }
  }

  # Backtrack through the table. This function will
  # be defined shortly.
  backtrack_matrix_mult(1, n, dims, N, matrices)
}
```

We use a table of matrix dimensions because it allows us to compute the minimum of the expression using a vector expression over k, something we could not do using the A list quite as easily. We loop over the length of intervals rather than just i and j because we need to compute the N[i,j] values in order of increasing lengths for the dynamic programming algorithm to work. If we did not do it in this order, we would not be guaranteed that the N values we use in the expression are filled out yet. Otherwise, though, we just implement the computation sketched earlier.

The backtracking function is equally simple. We want to find the optimal way of multiplying matrices i to j, and we have the table that tells us what N_{ij} is. So, we should find a split point where we can get that value from the recursion. That is where we should set the parentheses and then solve to the left and right, recursively, until we get to the base case of a single matrix, which of course is already the result we should return.

```r
backtrack_matrix_mult <- function(i, j, dims, N, matrices) {
  if (i == j) {
    matrices[[i]]
  } else {
    k <- i:(j - 1)
    candidates <- dims[i,1]*dims[k,2]*dims[j,2] + N[i,k] +
    N[k + 1,j]
    split <- k[which(N[i,j] == candidates)][1]
    left <- backtrack_matrix_mult(i, split, dims, N, matrices)
    right <- backtrack_matrix_mult(split + 1, j, dims, N,
    matrices)
    matrix_mult(left, right)
  }
}
```

At each step in the backtracking function, we construct a multiplication object using matrix_mult, so we rearrange the original expression in this way.

Expression Rewriting

With the dynamic programming algorithm in place, we know how to arrange multiplications in optimal order. We need to have them in a list, however, to access them by index in constant time in the backtracking function, but what we have as input is an expression that gives us a tree of mixed multiplications, addition, and data objects. So, the first step we must perform in the rearrangement is to collect the components of the multiplication in a list.

It is simple enough to visit all the relevant values in an expression. We recurse on all `matrix_mult` objects but not data or `matrix_sum` objects since it is these that we want to collect. It is inefficient to traverse the tree and grow an actual `list` object one element at a time; whenever you extend the length of a `list` object by one element, you need to copy all the old elements. Instead, we can implement a linked list—that we can prepend elements to in constant time—and translate that into a `list` object later.

To see this in action, we can consider a simpler tree first.

```
leaf <- function(x) structure(x, class = c("leaf", "tree"))
inner <- function(left, right)
  structure(list(left = left, right = right),
            class = c("inner", "tree"))
```

Let's say we have such a tree as shown here (and we do not *a priori* know that it has four leaves):

```
tree <- inner(leaf(1), inner(inner(leaf(2), leaf(3)), leaf(4)))
```

And say we want to construct a list containing the values in the leaves.

One way to implement linked lists is as a `list` object containing two values, which are the head of the list (an actual value) and the tail of the list (another list, or potentially `NULL` representing the empty list).

Since head and `tail` are useful built-in functions in R, we will call these two elements `car` and `cdr` instead. These are the names they have in the Lisp programming language and many other functional programming languages. We can construct a list from a `car` and `cdr` element like this:

```
cons <- function(car, cdr) list(car = car, cdr = cdr)
```

To traverse a tree, we use recursion, but we don't want to test the class of subtrees explicitly. Here is what we would usually do: have a test for the base case of having a leaf and another case for when we have an inner node. This approach would be sensible on this simple tree. However, once we start working with expressions where we can have many different node types, it might not be obvious what should be considered a base case or a recursive case. For any particular traversal, it is better to use generic functions.

```
collect_leaves_rec <- function(tree, lst)
  UseMethod("collect_leaves_rec")

collect_leaves_rec.leaf <- function(tree, lst) {
  cons(tree, lst)
}
collect_leaves_rec.inner <- function(tree, lst) {
  collect_leaves_rec(tree$left, collect_leaves_rec(tree$right,
  lst))
}
```

Using a generic function like this is certainly overkill for this simple example, but it illustrates the idea, which will be useful for more complex trees. Each node type is responsible for handling itself and potentially recursing further if this is needed. Here, the leaf handler prepends the tree to the list that is passed down the recursion. The tree is just the leaf, so this is the value we want to collect. The result is an updated list that we return from the recursion. For inner nodes, we first call recursively toward the right, passing along the lst object. This will prepend the elements in the right subtree to create a new list that we then pass along to a recursion on the left subtree.

The result of this traversal is a linked list containing all the leaves. To create a list object out of this, we need to run through the list and compute its length, allocate a list of that length, and then run through the linked list again to insert the elements in the list. This is one of the few tasks in R that is easier done with a loop than a functional solution, so that is what we will use.

```
lst_length <- function(lst) {
  len <- 0
  while (!is.null(lst)) {
    lst <- lst$cdr
    len <- len + 1
  }
  Len
}
lst_to_list <- function(lst) {
  v <- vector(mode = "list", length = lst_length(lst))
  index <- 1
  while (!is.null(lst)) {
    v[[index]] <- lst$car
    lst <- lst$cdr
    index <- index + 1
  }
  v
}
```

To improve the readability of the example, we will just add a function that gives us a vector instead of a list.

```
lst_to_vec <- function(lst) unlist(lst_to_list(lst))
```

Now we can use the combination of the traversal and transformation from the linked list to implement the function we want.

```
collect_leaves <- function(tree) {
  lst_to_vec(collect_leaves_rec(tree, NULL))
}
collect_leaves(tree)
```

```
## [1] 1 2 3 4
```

We can use the same approach to implement a better version of the rearrange_matrix_expr.matrix_mult function from earlier—one that rearranges the multiplication instead of just returning the original expression. We need it to collect the components of the multiplication—those would be data and sum objects—and then rearrange them using the dynamic programming algorithm.

```
rearrange_matrix_expr.matrix_mult <- function(expr) {
  matrices <- collect_mult_components(expr)
  arrange_optimal_matrix_mult(matrices)
}
```

The collect_mult_components function can be implemented using a traversal with a generic function like this:

```
collect_mult_components_rec <- function(expr, lst)
  UseMethod("collect_mult_components_rec")
collect_mult_components_rec.default <- function(expr, lst)
  cons(rearrange_matrix_expr(expr), lst)

collect_mult_components_rec.matrix_mult <- function(expr, lst)
    collect_mult_components_rec(expr$left,
             collect_mult_components_rec(expr$right, lst))

collect_mult_components <- function(expr)
    lst_to_list(collect_mult_components_rec(expr, NULL))
```

We use the default implementation to prepend expressions that are not multiplications to the list we are building, while we call recursively on the multiplication objects. Once we have collected all the components we need in a linked list, we translate it into a `list` object that lets us look up elements by index, as we need in the dynamic programming algorithm.

To see the rearranging in action, we can create the expression we used in the previous chapter. We have four matrices that we multiply together without setting any parentheses.

```
A <- matrix(1, nrow = 400, ncol = 300)
B <- matrix(1, nrow = 300, ncol = 30)
C <- matrix(1, nrow = 30, ncol = 500)
D <- matrix(1, nrow = 500, ncol = 400)

expr <- m(A) * m(B) * m(C) * m(D)
```

This implicitly sets parentheses such that the expression will be evaluated by multiplying from left to right.

```
expr
```

```
## (((([A] * [B]) * [C]) * [D])
```

This, however, is not the optimal order. Instead, it is better to first multiply A with B and C with D and then multiply the results.

```
rearrange_matrix_expr(expr)
```

```
## (([A] * [B]) * ([C] * [D]))
```

Expression Evaluation

We want to do more than manipulate matrix expressions; we want to evaluate them. This is something we can do easily in a recursive way, using a generic function to handle the different cases once again.

```
eval_matrix_expr <- function(expr) UseMethod("eval_matrix_expr")
eval_matrix_expr.matrix_data <- function(expr) expr$data
eval_matrix_expr.matrix_mult <- function(expr)
  eval_matrix_expr(expr$left) %*% eval_matrix_expr(expr$right)
eval_matrix_expr.matrix_sum <- function(expr)
  eval_matrix_expr(expr$left) + eval_matrix_expr(expr$right)
```

The base case, the `matrix_data` case, gives us an R object that should be a matrix. In the recursive calls, we use matrix multiplication (%*%) and addition (+) on the results of recursive calls, so what we apply to these operators on are R objects—which means that the + operator is *not* the operator we wrote to create `matrix_sum` objects.

Since we are explicitly delaying the evaluation of matrix expressions so we can rearrange them for optimal evaluation, we need a way to trigger the actual evaluation, and this would be the natural place to rearrange an expression as well, so we write a function for that.

```
v <- function(expr) eval_matrix_expr(rearrange_matrix_expr(expr))
```

Of course, we can also combine the parsing—the meta-programming approach that we looked at earlier—and an evaluation of the expression. We can write a function for doing faster evaluation of an expression like this:

```
fast <- function(expr) {
  v(build_matrix_expr(substitute(expr), parent.frame()))
}
```

As long as we stick to %*% and + operators, this function will evaluate to the same value as a plain matrix expression.

```
all(A %*% B %*% C %*% D == fast(A %*% B %*% C %*% D))
```

```
## [1] TRUE
```

28

However, it is not generally usable because we have changed the definition of *. You can modify the parser, though, and you have an optimizer for speeding up your matrix multiplications.

```
res <- microbenchmark(A %*% B %*% C %*% D,
                      fast(A %*% B %*% C %*% D))
options(microbenchmark.unit="relative")
print(res, signif = 3, order = "mean")

## Unit: relative
##                          expr min   lq mean median
##   fast(A %*% B %*% C %*% D) 1.0 1.00 1.00   1.00
##        A %*% B %*% C %*% D 5.8 5.79 5.09   5.85
##    uq  max neval
## 1.00 1.00   100
## 5.41 2.36   100
```

The recursion in build_matrix_expr stops the first time it does not recognize a call object and creates a data object. A better implementation would try to go deeper and optimize as much of the expression as it could, but this is more an exercise in meta-programming than in domain-specific languages.

As a DSL, matrix algebra is really simple. It's so simple that you might not consider it a language at all perhaps, but it is; algebraic notation is a DSL that is so useful that we get so familiar with it that we forget how amazing it is compared to the alternative—prose. Still, what we have implemented in this chapter *is* very simple, and while we might use the meta-programming techniques for code optimization, we probably would not write a DSL for something as simple as this. Nevertheless, the example illustrates the phases in reading, analyzing, and evaluating expressions we see in most DSLs. The three phases can be simpler or more complex in other DSLs (the "analysis" step might be entirely missing), and they might be merged so parsing and evaluation are done as a single step, but conceptually these are the steps we usually see.

CHAPTER 3

Components of a Programming Language

While this is not a book about compilers and computer languages in general, it will be helpful to have some basic understanding of the components of software that parse and manipulate computer languages—or at least domain-specific computer languages.

When we write software for processing languages, we usually structure this such that the input gets manipulated in distinct phases from the raw input text to the desired result, a result that is often running code or some desired data structures. When processing an embedded DSL, however, there is not necessarily a clear separation between parsing your DSL, manipulating expressions in the language, and evaluating them. In many cases, embedded DSLs describe a series of operations that should be executed sequentially—this is, for example, the case with graphical transformations in `ggplot2` or data manipulation in `magrittr` and `dplyr`. When this is the case, you wouldn't necessarily split evaluations of DSL expressions into a parsing phase and an evaluation phase; you can perform transformations one at a time as they are seen by the R parser. Conceptually, though, there are still two steps involved—parsing a DSL statement and evaluating it—and you have to be explicit about this with

© Thomas Mailund 2018
T. Mailund, *Domain-Specific Languages in R*,
https://doi.org/10.1007/978-1-4842-3588-1_3

more complex DSLs. Even with simple DSLs, however, there are benefits to keeping the different processing phases separate. It introduces some overhead in programming as you need to represent the input language in some explicit form before you can implement its semantics, but it also allows you to separate the responsibility of the various processing phases into separate software modules, making them easier to implement and test.

This chapter describes the various components of computer languages and the phases involved in processing a domain-specific language. We will use the following packages:

```
library(magrittr) # for the %>% operator
library(Matrix)   # for the `exam` function
```

Text, Tokens, Grammars, and Semantics

First, we need to define some terminology. Since this book is not about language or parser theory, I will stick with a few informal working definitions for terms we will need later.

When we consider a language, we can look at it at different levels of detail, from the most basic components to the meanings associated with expressions and statements. For a spoken language, the most basic elements are the *phonemes*—the distinct sounds used in the language. When strung together, phonemes become words, words combine to make sentences, and sentences have meanings. For a written language, the atomic elements are *glyphs*—the letters in languages that are written using alphabets, such as English. Sequences of letters can form words, but a written sentence contains more than just words—we have punctuation symbols as well. Together, we can call these *tokens*. A string of *tokens* forms a sentence, and again, we assign meanings to sentences.

For computer languages, we have the same levels of abstractions on strings of symbols. The most primitive level is just a stream of input characters, but we will have rules for translating such character sequences into sequences of *tokens*. This process is called *tokenization*. The formal definition of a programming language will specify what the available tokens in the language are and how a string of characters should be translated into a string of tokens.

Consider the following string of R code:

foo(x, 2*x)

This is obviously a function call, but seen by the tokenizer it is a string of characters that it needs to translate into a sequence of tokens. It will produce this:

```
identifier["foo"] '(' identifier["x"],
                number[2], '*', identifier["x"]
              ')'
```

I am using a home-brewed notation for this, but the idea is that a tokenizer will recognize that there are some identifiers and numbers (and it will identify what those identifiers and numbers are) and then some verbatim tokens such as '(', '*', and ')'.

The tokenizer, however, will be equally happy to process a string such as this:

foo **x** (2) x *

into the following sequence:

```
identifier["foo"] identifier["x"] '('
                number[2] ')' identifer["x"] '*'
```

This is obviously not a valid piece of R code, but the tokenizer does not worry about this. It merely translates the string into a sequence of tokens (with some associated data, such as the strings "foo" and "x" for the identifiers and the number 2 for the number). It does not worry about higher levels of the language.

When it comes to tokenizing an embedded language, we are bound to what that language will consider as valid tokens. We cannot create arbitrary kinds of tokens since all languages we write as embedded DSLs must also be valid R. The tokens we can use are either already R tokens or variables and functions we define to have special meaning. Mostly, this means creating objects through function calls and defining functions for operator overloading.

What a language considers a valid string of tokens is defined by its *grammar*.[1] A *parser* is responsible for translating a sequence of tokens into an expression or a language statement. Usually, what a parser does is translate a string of tokens into an expression *tree*—often referred to as an *abstract syntax tree* (AST).[2] The tree structure associates more structure to a piece of code than the simple sequential structure of the raw input and the result of the tokenization. An example of how an abstract syntax tree for the function call we tokenized earlier could look like Figure 3-1. Here, the italic labels refer to a syntactic concept in the grammar, while the monospace font labels refer to verbatim input text. Tokens are shown in gray boxes. As we see, these can either be verbatim text or have some grammatical information associated with them, describing what type of token they are (in this example, this is either an identifier or a number).

---

[1]Technically, what I refer to as *grammar* is *syntax*. Linguists use *grammar* to refer to both *morphology* and *syntax*, where *syntax* is the rules for stringing words together. In computer science, though, the term *grammar* is used as I use it here. Therefore, I will use *syntax* and *grammar* interchangeably.

[2]The purists might complain here and say that a parser will construct a *parse tree* and not an AST. The difference between the two is that a parse tree contains all the information in the input (such as parentheses, spaces, and so on) but not the meta-information about which grammatical structures it represents. The AST contains only the relevant parts of the input but includes grammatical information on top of that. If you want, you can consider first parsing and then translating the result into an AST as two separate steps in handling an input language. I consider them part of the same and will claim that a parser constructs an AST.

When there is information associated, I have chosen to show this as two nodes in the tree: one that describes the syntactical class the token is (identifier or number) and a child of that node that contains the actual information (foo, x, and 2 in this case).

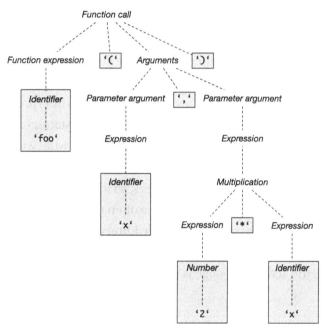

Figure 3-1. *Example of an abstract syntax tree for a concrete function call*

Grammatical statements are those a parser will consider valid. Such sentences are, if we return to natural languages, those sentences that follow the grammatical rules. This set of grammatical sentences is distinct from the set of sentences that have some associated *meaning*. It is entirely possible to construct meaningless but grammatically correct sentences. The sentence "Colourless green ideas sleep furiously" is such a sentence, created by the linguist Noam Chomsky. It is entirely grammatical and also completely meaningless. *Semantics* is the term we use to link grammatical sentences to their meaning. You will know this distinction in programming languages when you run into runtime exceptions. If you get an exception when you run a program, you will have constructed a grammatical sentence—otherwise, the parser would have complained about syntactical errors—but the sentence is violating language rules in some other way. A grammatical sentence does not need to have a well-defined meaning. For example, an expression that adds two variables, x + y, is grammatical, but if the variables contain two incompatible types (x could be a string while y is a number), the sentence violates the language's type rules. Semantics, when it comes to programming languages, define what actual computations a statement describes. A compiler or an interpreter—the latter for R programs—gives meaning to grammatical statements.[3]

For embedded DSLs, the semantics of a program is what we do to evaluate an expression once we have parsed it. We are not going to formally specify semantics or implement interpreters, so for this book, the semantics part of a DSL is plain old R programs. More often than not, what we use embedded DSLs for is an interface to some library or framework. It is the functionality of this framework that provides the semantics of what we do with the DSL; the actual language is just an interface to the framework.

Specifying a Grammar

Since we are using R to parse expressions, we do not have much flexibility in what can be considered tokens, and we have some limitations in the kinds of grammars we can implement. However, we have some flexibility for the grammars in how we combine functions and operators.

---

[3]Notice, however, that there is a distinction between giving a statement meaning and giving it the *correct* meaning. Just because your program computes *something* doesn't mean it computes what you intended it to compute. When we construct a language, domain-specific or general, we can give meaning to statements, but we cannot—this is theoretically impossible—guarantee that it is the *intended* meaning. That will always be the responsibility of the programmer.

To specify grammars in this book, I will take a traditional approach and describe them in terms of "rules" for generating sentences valid within a grammar. Consider the following grammar:

```
EXPRESSION ::= NUMBER
            |  EXPRESSION '+' EXPRESSION
            |  EXPRESSION '*' EXPRESSION
            |  '(' EXPRESSION ')'
```

This grammar describes rules for generating expressions consisting of the addition and multiplication of numbers, with parentheses to group expressions.

You should read this as "An expression is either a number, the sum of two expressions, the product of two expressions, or an expression in parentheses." The definition is recursive—an expression is defined in terms of other expressions—but we have a base case, a number, that lets us create a base expression, and from such an expression we can generate more complex expressions.

The syntax I use here for specifying grammars is itself a grammar—a meta-grammar if you will. The way you should interpret it is thus: the grammatical object we are defining is to the left of the ::= object. After that, we have a sequence of one or more ways to construct such an object, separated by |. These rules for constructing the object we define will be a sequence of other grammatical objects. These can be either objects we define by other rules (I will write those in all capitals and refer to them as *meta-variables*) or concrete lexical tokens (I write those in single quotes, as with the '+' in the second rule for creating a sum). This notation contains the same information as the graphical notation I used in Figure 3-1 where meta-variables are shown in italics and concrete tokens are put in gray boxes.

Meta-grammars like this are used to define languages formally, and there are many tools that will let you automatically create parsers from a grammar specification in a meta-grammar. I will use this home-made meta-grammar much less formally. I use it as a concise way to describe the grammar of DSLs we create or simply a pseudo-code for grammar.

To create an expression, we must follow the meta-grammar rules by using one of the four alternatives provided: either reduce an expression to a number (a sum or product) or create another in parentheses. For example, we can apply the rules in turn and get the following:

```
EXPRESSION > EXPRESSION '*' EXPRESSION                        (3)
           > '(' EXPRESSION ')' '*' EXPRESSION                (4)
           > '(' EXPRESSION '+' EXPRESSION ')' '*' EXPRESSION (2)
           > '(' number[2] '+' number[2] ')' '*' number[3]  (1x3)
```

This lets us construct the expression `(2 + 2) * 3` from the rules.

If there are different ways to go from meta-variables to the same sequence of terminal rules (there are rules that lead to the exact sequence of lexical tokens), then we have a problem with interpreting the language. The same sequence of tokens could be interpreted as specifying two different grammatical objects. For the expression grammar, we have ambiguities when we have expressions such as `2 + 2 * 3`. We can parse this in two different ways, depending on which rules we apply to get from the meta-variable `EXPRESSION` to the concrete expression. We can apply multiplication first and get what amounts to `(2 + 2) * 3`, or we can apply the addition rule first and get `2 + (2 * 3)`. We know from the traditional mathematical notation that we should get the second expression because multiplication has higher precedence[4] than addition, so the * symbol binds 2 and 3 together tighter than + does 2 and 2, but the grammar does not guarantee this. The grammar is ambiguous.

It is possible to fix this by changing the grammar to this:

```
EXPRESSION ::= TERM '+' EXPRESSION | TERM
TERM ::= TERM '*' FACTOR | FACTOR
FACTOR ::= '(' EXPR ')' | NUMBER
```

This is a more complex grammar that lets you create the same expressions, but through three meta-variables that are recursively defined in terms of each other. It is structured such that products will naturally group closer than sums—the only way to construct the expression 2 + 2 * 3 is the parse tree shown in Figure 3-2. The order in which we apply the rules can vary, but the tree will always be this form and group the product closer than the sum.

---

[4]*Operator precedence* is a term we use to describe how "tightly" different operators bind, i.e., which operators are invoked before others, or put in another way, where we implicitly set parentheses in an expression. Multiplication binds tighter than addition—has a higher precedence—so the expression 2*x + 4*y is interpreted as (2*x) + (4*y) rather than, for example, 2*(x + (4*y)). Precedence also tells us whether operators are evaluated left to right or right to left, so x + y + z is evaluated as (x + y) + z rather than x + (y + z) because + evaluates left to right. For addition, the left-to-right or right-to-left order doesn't matter because we will end up with the same result, but for some operators it does, for example, exponentiation, which is evaluated left to right. So, x**y**z is x**(y**z), which (usually) is different from (x**y)**z. If you use operators in a domain-specific language, the precedence of the operators will affect your language.

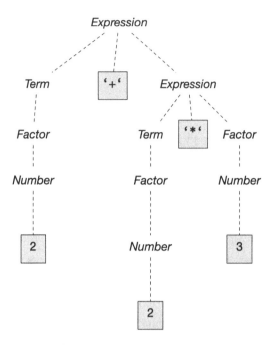

Figure 3-2. *Parse tree for 2 + 2 * 3*

An unambiguous grammar is preferable over an ambiguous one for obvious reasons, but creating one can complicate the specification of the grammar, as we see for expressions. This can be alleviated by making a smarter parser that takes such things as operator precedence into account or keeps track of context when parsing a string. Regardless of whether we write smarter parsers or unambiguous grammars, we would never work long with expression trees as complex as that shown in Figure 3-2—this tree explicitly shows all grammar meta-variables, but in practice, we would simplify it after parsing it and before processing the expression.

When writing embedded DSLs, we are stuck with R's parser, and we must obey its rules. If you are writing your own parser entirely, you can pass context along as you parse a sequence of tokens, but if you want to exploit R's parser and create an embedded DSL, you are better off ensuring that all grammatically valid sequences of tokens unambiguously refer to one grammatical meta-variable. Precedence ambiguities will be taken care of by R as will associativity—the rules mean that 1 + 2 + 3 + 4 is interpreted as (((1 + 2) + 3) + 4). Exploiting R's parsing rules will allows us to construct languages where each expression uniquely matches a parser meta-variable if we are a little careful with designing the language.

As an example grammar that isn't just expressions, imagine that we want a language for specifying graphs—graphs as in networks or state machines, not plots. We can define a grammar for *directed acyclic graphs* (DAGs) by saying that a DAG is either an empty graph or a graph followed by an edge.

```
DAG ::= 'dag()' | DAG '+' EDGE
```

We use a function, dag(), to create an empty DAG. Calling this function brings us into the graph specification DSL and gives us an object we can use to program the grammar operations in R. We will use the plus operator to add edges to a DAG. That is a somewhat arbitrary choice, but it makes it easy to implement the parser since we simply will have to overload the generic + function. For edges, we will keep it simple and just require that we have a "from" node and a "to" node. We can define our own infix operator to create them.

```
EDGE ::= NODE '%=>%' NODE
```

We cannot define any infix operator we want—we would be out of luck, for example, if we wanted the operator to be ==> since R's parser would interpret that as two tokens, == and >. We can always define our own, however, if we name them something starting and ending with the percentage sign. We can also reuse the existing infix operators through overloading, as we will do with the plus sign to add edges to a DAG, but for this graph grammar, we can run into some problems if we attempt this, as we will see next. For nodes, we will not expand them more now. They are atomic tokens—we can, for example, require them to be strings.[5]

---

[5]We haven't formally defined how we would specify non-literate tokens in the syntax we use for specifying grammars, and doing so will not make the example any clearer, so let's just state that informally.

We will dig more into writing parsers in the next chapter, but for this simple language we can quickly create one. The parser needs to collect edges, so we will use linked lists for this, and it is natural to make an empty DAG contain an empty list of edges. Then, adding edges to a DAG means prepending them to this list. Such an implementation is as simple as this:

```r
cons <- function(car, cdr) list(car = car, cdr = cdr)
dag <- function() structure(list(edges = NULL), class = dag")
`%=>%` <- function(from, to) c(from, to)
`+.dag` <- function(dag, edge) {
    dag$edges <- cons(edge, dag$edges)
    dag
}
```

With only these four functions, we can create a DAG using syntax like this:

```r
dag() +
  "foo" %=>% "bar" +
  "bar" %=>% "baz"
```

It might not be the best syntax we can come up with, but it's easier to read than nested function calls.

```r
add_edge <- function(dag, from, to) {
  dag$edges <- cons(c(from, to), dag$edges)
  dag
}
add_edge(add_edge(dag(), "foo", "bar"), "bar", "baz")
```

Using the pipe operator from magrittr might be even more readable, though, for people familiar with it.

```r
dag() %>% add_edge("foo", "bar") %>% add_edge("bar", "baz")
```

In any case, we have built a small language that we can parse by defining only four functions—three if we discount the list cons function, which isn't specific to the language.

We used %=>% to construct edges. Could we use => instead? The short answer is no. R's parser will consider this two tokens, = and >, and although we *could* define a function with that name, using backquotes to make it a valid identifier, we wouldn't get an infix operator.

```
`=>` <- function(from, to) c(from, to)
`=>`("foo", "bar")
"foo" => "bar"

## Error: <text>:3:8: unexpected '>'
## 2: `=>`("foo", "bar")
## 3: "foo" =>
##                ^
```

If we want to have an infix operator that does not use percentage signs, we have to overload one of the operators that R already has—and => is not one of them (greater-than-or-equal is >=).

Could we use > instead then? This is an R infix operator, and, therefore, we can overload it. We just need a type for a node to do this. If we keep nodes specified as strings, we would have to change the string operator, and we do not want to do that—it could potentially break a *lot* of existing code—so the best approach would be to define a node class to work with.

```
node <- function(name) structure(name, class = "node")
`>.node` <- function(from, to) c(from, to)
```

With these functions, we can create an edge with this syntax:

```
node("foo") > node("bar")

## [1] "foo" "bar"
```

Changing a % infix operator to > changes the precedence, however. A % operator has higher precedence than +, which is why we got edges that we could add to the DAG earlier, but > has lower precedence than +, so we add the left node to the DAG first and only then invoke the > operator.

```
dag() + node("foo") > node("bar")
```

```
## $edges
## $edges$car
## [1] "foo"
## attr(,"class")
## [1] "node"
##
## $edges$cdr
## NULL
##
##
## [[2]]
## [1] "bar"
```

We can fix this using parentheses, of course.

```
dag() + (node("foo") > node("bar"))
```

```
## $edges
## $edges$car
## [1] "foo" "bar"
##
## $edges$cdr
## NULL
##
##
## attr(,"class")
## [1] "dag"
```

It is not particularly safe to rely on programmers to remember parentheses, so a better solution is to get the precedence right. We can do that by choosing a different operator for adding edges to DAGs. If we replace + with |, for example, we get the right behavior since | has lower precedence than >.

```
`|.dag` <- function(dag, edge) {
  dag$edges <- cons(edge, dag$edges)
  dag
}
dag() | node("foo") > node("bar")

## $edges
## $edges$car
## [1] "foo" "bar"
##
## $edges$cdr
## NULL
##
##
## attr(,"class")
## [1] "dag"
```

There are pros and cons to using operator overloading. Having to change string tokens into node tokens requires more typing, but on the other hand, we can use this to validate expressions while we parse them and make sure that nodes are strings.

Alternatively, we could use meta-programming and explicitly traverse expressions to make sure that the > operator will be the edge-creating operator instead of string comparison, similar to how we rewrote matrix expressions in the previous chapter.

Returning to the `magrittr` solution for a brief moment, I think it is worth mentioning that designing a language is not all about defining new syntax. The language we are defining here, for specifying graphs, is doing the same as the pipe operator does, so in this particular case, we do not *need* to specify a new grammar to get all the benefits we want to achieve. When using pipes, we avoid the nested function calls that would make our code hard to read, and we can specify a DAG as a list of edges that we add to it. The pipe operator will be familiar to most programmers, and best of all, if we use it, we do not need to implement any parsing code. We are still creating a DSL, though, when we define the functions to manipulate a DAG. Providing functions that give you a vocabulary to express domain ideas is also language design. The `dplyr` package is an example of this— it is used together with the pipe operator to string various operations together, so it does not provide much in terms of new syntax, but it provides a very strong language for specifying data manipulation.

Of the various solutions we have explored, I prefer the pipe-based. It makes it easy to extend edge information to more than a "from" node and a "to" node—which is hard with a binary operator—and we can implement it without any language code; we just have to make the DAG the first argument to all the manipulation functions we would add to the language. Of course, this solution is possible only because the language we considered was a simple string of operations. This is not always the case, so sometimes we do need to do a bit more work.

Designing Semantics

The reason we write domain-specific languages is to achieve some effect— we want to associate meaning, or semantics, to expressions in the DSL, and we want our DSL expressions to achieve some result, whether that is by executing some computations or by building some data structures. The purpose of a DSL merely is to provide a convenient interface to whatever semantics we want the language to have.

If we always make our parsing code construct a parse tree, then the next step in processing the DSL involves manipulating this tree to achieve the desired effect. There might be several steps involved in this—for example, we rewrote expressions in the matrix expression example to optimize computations—but at the end of the processing we will execute the commands the DSL expression describes.

Executing the DSL is sometimes straightforward and can be done as a final traversal of the parse tree. This is what we did with the matrix expressions where the purpose of the DSL was to rewrite expressions rather than evaluate them—the latter being a simple matter of multiplying and adding matrices. In other cases, it makes sense to separate the semantic model and the DSL by having a framework for the actions we want the language to execute. Having a framework for the semantics of the language lets us develop and test the semantic model separately from the language we use as an interface for it; it even allows us to have different input languages for manipulating the same semantic model—not that I recommend having many different languages to achieve the same goals.

To illustrate this, we can consider a language for specifying a finite state continuous time Markov chain (CTMC).[6] I chose this example because we have already implemented several versions of a finite state system when we implemented the graph DSL, and for a CTMC we just

---

[6]If you are not familiar with *continuous time Markov chains* an informal description could be this: A CTMC is a graph where nodes represent states and edges transition between states, and each edge is annotated with the rate at which changes happen. The model is stochastic, so rates should be thought of as expectations of change over time. If we have a rate of 1 between states s and t, then for each time unit, we would expect to see one change from s to t. If we want to know the probability of being in state t at some time τ if we are in state s at time =0, we cannot consider the states s and t in isolation. There might be other states we can go to from s and if we go to those, it might take a longer or shorter time to reach t, or there might be states we can continue to from t so we might go *through* t before time τ but no longer be there. We have to take all edges and all rates into account. Figuring out the probability of being at state t at time τ is therefore dependent on matrix operations, in particular matrix exponentiation.

have to associate rates with all the edges. Continuous time Markov chains are used many places in mathematical modeling, and the implementation mostly comes down to specifying an instantaneous rate matrix. This is a matrix that specifies the rate at which we move from one state to another. Such a matrix should have non-negative values on all off-diagonal entries, and on the diagonal, we should have minus the sum of the other entries in the rows. In a framework where we use CTMCs, we would likely implement the functionality to work on rate matrices, but for specifying the CTMCs, a domain-specific language might be easier to use.

Calling it the "semantics" of the language to translate graph specifications into a matrix might be stretching the word, but if we consider the DSL as a way of specifying models and the (imagined) framework that manipulates them as part of the language, then I think we can justify it. Using the language will consist of specifying the CTMC, translating it into the corresponding rate matrix, and then manipulating it as intended. The language part of it is the translation from the specification into a matrix.

As for the graphs shown previously, we need to specify the edges in the chain. We need to have a rate associated with each edge, so the most natural syntax will be the pipe version—with this version it is simpler to specify three values for an edge: the "from" and "to" states and the rate. I will keep these in three different linked lists, just because it makes it easier to construct the matrix this way.

We reuse these two functions we wrote in the previous chapter:

```
cons <- function(car, cdr) list(car = car, cdr = cdr)
lst_length <- function(lst) {
  len <- 0
  while (!is.null(lst)) {
    lst <- lst$cdr
    len <- len + 1
  }
```

```r
    len
}
lst_to_list <- function(lst) {
  v <- vector(mode = "list", length = lst_length(lst))
  index <- 1
  while (!is.null(lst)) {
    v[[index]] <- lst$car
    lst <- lst$cdr
    index <- index + 1
  }
  v
}
```

We then define the following functions for specifying CTMCs:

```r
ctmc <- function()
  structure(list(from = NULL,
                 rate = NULL,
                 to = NULL),
            class = "ctmc")

add_edge <- function(ctmc, from, rate, to) {
  ctmc$from <- cons(from, ctmc$from)
  ctmc$rate <- cons(rate, ctmc$rate)
  ctmc$to <- cons(to, ctmc$to)
  ctmc
}
```

Translating the lists into a rate matrix is now simply a normal programming job. We collect the nodes from the "from" and "to" lists—we translate them into R lists first since those are easier to work with once we are done collecting elements—and we then get the unique node names.

These become the rows and columns of the rate matrix, and we iterate through all the edges to insert the rates. After that, we set the diagonal, and we are done.

```
rate_matrix <- function(ctmc) {
  from <- lst_to_list(ctmc$from)
  to <- lst_to_list(ctmc$to)
  rate <- lst_to_list(ctmc$rate)
  nodes <- c(from, to) %>% unique %>% unlist

  n <- length(nodes)
  Q <- matrix(0, nrow = n, ncol = n)
  rownames(Q) <- colnames(Q) <- nodes

  for (i in seq_along(from)) {
    Q[from[[i]], to[[i]]] <- rate[[i]]
  }

  diag(Q) <- - rowSums(Q)

  Q
}
```

Constructing a CTMC rate matrix using this small language is now as simple as this:

```
Q <- ctmc() %>%
  add_edge("foo", 1, "bar") %>%
  add_edge("foo", 2, "baz") %>%
  add_edge("bar", 2, "baz") %>%
  rate_matrix()
Q
```

```
##      bar foo baz
## bar   -2   0   2
## foo    1  -3   2
## baz    0   0   0
```

Once we have translated the CTMC into this matrix, we can consider the language design over. However, integrating CTMC construction and the operations that we can do on a CTMC will be important for ease of use of the DSL and can be considered part of the language as well. In this particular case, the good news is that the actual language design is done for us. The pipe operator tells us how to combine our CTMCs with further processing—we just have to write functions that can be used in a pipe. For example, if we want to know the transition probabilities of the CTMC over a time interval—i.e., we want to know the probability of going from any one state to any other over a given time—we can add a function for that. The probabilities can be computed using matrix exponentiation (if you are not familiar with CTMC theory, just trust me on this). To make such a function compatible with a pipeline, we simply have to make the most likely data to come from the left in a pipe the first argument of the function. So, we could write this:

```
transitions_over_time <- function(Q, t) expm(Q * t)
P <- Q %>% transitions_over_time(0.2)
P
```

```
## 3 x 3 Matrix of class "dgeMatrix"
##              bar       foo      baz
## bar 0.6703200 0.0000000 0.32968
## foo 0.1215084 0.5488116 0.32968
## baz 0.0000000 0.0000000 1.00000
```

Even with the language constructions in place—the pipe operator—there is still some language design to be done. It is not always obvious what the flow of data will be through a pipe after all. An example is when we want to evolve vectors of state probability.

```
probs <- c(foo=0.1, bar=0.9, baz=0.0)
```

Is it more natural to have the probability vectors flow through the pipeline?

```
evolve <- function(probs, Q, t) {
  probs <- probs[rownames(Q)]
  probs %*%transitions_over_time(Q, t)
}
probs %>% evolve(Q, 0.2)
```

```
## 1 x 3 Matrix of class "dgeMatrix"
##             bar        foo      baz
## [1,] 0.6154389 0.05488116 0.32968
```

Or, would it be more natural to always have the CTMC (or its rate matrix) flow through the pipeline?

```
evolve <- function(Q, t, probs) {
  probs <- probs[rownames(Q)]
  probs %*%transitions_over_time(Q, t)
}
Q %>% evolve(0.2, probs)
```

```
## 1 x 3 Matrix of class "dgeMatrix"
##             bar        foo      baz
## [1,] 0.6154389 0.05488116 0.32968
```

The former might feel more natural if we think of the system evolving over time, but the latter would fit better with a pipeline where we construct the CTMC first, then translate it into a rate matrix, and finally evolve the system.

```
ctmc() %>%
  add_edge("foo", 1, "bar") %>%
  add_edge("foo", 2, "baz") %>%
  add_edge("bar", 2, "baz") %>%
  rate_matrix() %>%
  evolve(0.2, probs)

## 1 x 3 Matrix of class "dgeMatrix"
##              bar       foo    baz
## [1,] 0.6154389 0.05488116 0.32968
```

Only use cases and experimentation will tell us what the best language design is. But then, that is also what makes designing languages so interesting.

## CHAPTER 4

Functions, Classes, and Operators

Everything you do in R, you do with functions. Consequently, if you want to implement a domain-specific language, you must do so by writing functions. All the actions that your new DSL should support must be implemented using functions, and should you want a special syntax for your DSL, you will have to write functions for parsing this syntax. When implementing an embedded DSL, as we shall see, much of the parsing can be outsourced to R's parser. The price for this is some restrictions to the syntax for the DSL—the DSL must be syntactically valid R code. We cannot construct arbitrary syntaxes for embedded languages, but by using operator overloading or meta-programming and by defining new infix operators, we do have some flexibility in designing our DSLs.

In the previous two chapters, we saw examples of how to use both operator overloading and meta-programming to treat R expressions as expressions in an embedded language. The purpose of this chapter is to go into the details of the operator option, while the next chapter will cover the possibility of explicitly manipulating expressions through meta-programming.

© Thomas Mailund 2018
T. Mailund, *Domain-Specific Languages in R*,
https://doi.org/10.1007/978-1-4842-3588-1_4

The S3 Object-Oriented Programming System

You could write a whole book about the object-oriented programming systems supported by R—I know because I have written such a book—but for operator overloading and implementing DSL parsers, we need only a few of the object orientation features, and this section will give you a quick introduction to those. We will use only the simplest object orientation system in this book, the S3 system. If you are already familiar with the S3 system, feel free to skip ahead to the next section.

Objects and Classes

The S3 system has a straightforward approach to object orientation. It lets us associate classes with any object (except for NULL). Classes are text strings with no structure, and you can get the classes associated with an object by calling the function class.

class(4)

[1] "numeric"

class("foo")

[1] "character"

class(TRUE)

[1] "logical"

class(sin)

[1] "function"

You can set the class of an object with the corresponding assignment function, shown here:

```
class(sin) <- "foo"
class(sin)
```

```
## [1] "foo"
```

without affecting what the object is or does.

```
sin(0)
```

```
## [1] 0
```

There are some limits to which objects you can change the class of—you cannot change the class of literal numbers and strings, for example, so if you attempted this:

```
class(2) <- "character"
class("foo") <- "numeric"
```

you would get errors. Still, you are free to modify the classes of objects any time you want. The class associated with an object is just an attribute containing one or more class names—in the previous cases, all the objects have a single class, but the matrix expressions we created Chapter 2, for example, had multiple classes. Vectors of class names are used both for multiple inheritance and for single inheritance, and there is no formal class structure at all. You can set the classes any way you want. The class attribute is just a vector of strings that are interpreted as class names.

The only thing that makes the class attribute interesting, compared to any other attributes you could associate with objects, is how it is used for dispatching generic functions. When you call a generic function, the actual function that gets called depends on the class of one of its arguments, usually the first (if you do not provide the argument explicitly, it will always be the first).

Generic Functions

The generic function mechanism in the S3 system is implemented in the UseMethod and NextMethod functions. To define a generic function, foo, you would, for example, write the following:

```
foo <- function(x, y, z) UseMethod("foo")
```

Calling foo would then invoke UseMethod that would search for concrete implementations of foo. Such functions are identified by their name alone—any function whose name starts with foo. is considered an implementation of foo, and the one that will be chosen depends on the class of the argument x. We haven't done *any* implementations of foo yet, so calling the function will just give us an error for now.

```
foo(1, 2, 3)
```

```
## Error in UseMethod("foo"): no applicable method for 'foo'
applied to an object of class "c('double', 'numeric')"
```

However, we can make a default implementation. The default implementation for a generic function—the one that will be used if UseMethod can find no better matching implementation—has a name that ends with .default. We can implement the following:

```
foo.default <- function(x, y, z) {
   cat("default foo\n")
}
```

and this function will be invoked if we call foo with an object that doesn't have a better implementation.

```
foo(1, 2, 3)
```

```
## default foo
```

To specialize a generic function to specific classes, we just have to define functions with appropriate names. Any function name that begins with foo. can be used and will be called if we call foo with an object of the appropriate class. To specialize foo to numeric values, for example, we could write the following:

```
foo.numeric <- function(x, y, z) {
    cat("numeric\n")
}
```

Now the following function will be called if x is numeric.

```
foo(1, 2, 3)
```

```
## numeric
```

A quick note before we continue to explore how dispatching is done on generic functions for user-defined classes: when we used UseMethod in the definition of foo, we called it with the name foo. This is why it looks for that name when it searches for implementations of the generic function. We could have asked it to search for other functions. The name we gave the generic function when we defined foo is not what determines what we search for when we call the function—that is determined by the name we give UseMethod. Further, we have seen it dispatches on the first argument of foo, but this is just a default. We *could* give UseMethod another object to dispatch on.

```
bar <- function(x, y, z) UseMethod("foo", y)
```

I do not recommend doing this—it goes against the conventions used in R—but it is possible, and with this function, we would dispatch on the y argument (and search for the generic function foo, not bar).

```
foo("foo",2,3)

## default foo

bar("foo",2,3)

## numeric

bar(1,"bar",3)

## default foo
```

Going back to the rules for dispatching, when UseMethod is called, it starts to search for functions based on their name. It will take the classes of an object and search those in order. If it doesn't find any matching function, it will call the default, if it exists.

So, let's consider again foo where we have a numeric and a default implementation.

```
x <- 1
foo(x, 2, 3)

## numeric
```

Here, we have created the object x that is numeric, so when we call foo, we match the numeric function. But we can change the class of x and see what happens.

```
class(x) <- c("a", "b", "c")
foo(x, 2, 3)

## default foo
```

Now, since x has the classes a, b, and c but not numeric, UseMethod does not find the numeric version but hits the default one. We can, of course, define functions for the other classes and see what happens.

```
foo.a <- function(x, y, z) cat("a\n")
foo.b <- function(x, y, z) cat("b\n")
foo.c <- function(x, y, z) cat("c\n")
foo(x, 2, 3)
```

```
## a
```

Because we now have functions for x's classes, we can find them, and because a is the first class, that is the one that will be called. If we change the order of x's classes, we call the other functions—UseMethod always calls the first it finds.

```
class(x) <- c("b", "a", "c")
foo(x, 2, 3)
```

```
## b
```

```
class(x) <- c("c", "b", "a")
foo(x, 2, 3)
```

```
## c
```

When calling UseMethod, we will find and call the first matching function. The related NextMethod function can be invoked to find and call the next function in the chain of classes. To see it in action, we can redefine the three a, b, and c foo implementations and make them call the next function in line.

```
foo.a <- function(x, y, z) {
  cat("a\n")
  NextMethod()
}
foo.b <- function(x, y, z) {
  cat("b\n")
  NextMethod()
}
```

```r
foo.c <- function(x, y, z) {
  cat("c\n")
  NextMethod()
}
```

They will all call the next function in line, and since we have a default implementation of foo, the last in line will call that one. The order in which the functions are called depends entirely on the order of x's classes.

```r
class(x) <- c("a", "b", "c")
foo(x, 2, 3)
```

```
## a
## b
## c
## default foo
```

```r
class(x) <- c("b", "a", "c")
foo(x, 2, 3)
```

```
## b
## a
## c
## default foo
```

```r
class(x) <- c("c", "b", "a")
foo(x, 2, 3)
```

```
## c
## b
## a
## default foo
```

This generic dispatch mechanism is extremely flexible, so it will require some discipline to ensure a robust object-oriented design. To implement domain-specific languages, though, we are interested in how we can use it to add operators to our language.

Operator Overloading

Most operators, with exceptions of arrow assignments (<- and ->) and slot and component access (@ and $), behave as generic functions and can be overloaded. The mechanism for overloading operators is not different from the mechanism for implementing a new version of a generic function— if you name a function the right way, it will be invoked when a generic function is called.

For example, to define addition on a objects, we need to define the +.a, which we could do like this:

```
`+.a` <- function(e1, e2) {
  cat("+.a\n")
  NextMethod()
}
x + 2

## +.a

## [1] 3
## attr(,"class")
## [1] "c" "b" "a"
```

Here, we print some output and then invoke the underlying numeric addition—since the object x here is a numeric value as well as an object of class a—by invoking NextMethod. It is important that we use NextMethod here. If we use addition, we would be calling +.a once again. So do not attempt this:

```
`+.a` <- function(e1, e2) {
  cat("+.a\n")
  e1 + e2
}
```

This function causes infinite recursion, and it is not what you want.

63

You overload operators in the same way as you would define specializations of generic functions, but there *are* differences between the two and how they dispatch. With standard generic functions, you would dispatch based on the first argument. With operators, there are heuristics that are there to make sure that the same function is called regardless of the order of the operands. You could easily imagine the pain to debug programs where switching the order of the operands would also call entirely different functions. That doesn't happen in R because operations are not *exactly* the same as other generic functions.

If we have defined +.a and we try to add a number to x, then we can do that in either order, and it will be the +.a function that is called.

```
x + 3
```

```
## +.a
```

```
## [1] 4
## attr(,"class")
## [1] "c" "b" "a"
```

```
3 + x
```

```
## +.a
```

```
## [1] 4
## attr(,"class")
## [1] "c" "b" "a"
```

This is also the case if we add to x an object of a different class.

```
x <- 1 ; y <- 3
class(x) <- "a"
class(y) <- "b"
x + y
```

```
## +.a
```

```
## [1] 4
## attr(,"class")
## [1] "a"
```

```
y + x
```

```
## +.a
```

```
## [1] 4
## attr(,"class")
## [1] "b"
```

unless we have also defined an addition operator for that class.

```r
`+.b` <- function(e1, e2) {
  cat("+.bn")
  NextMethod()
}
```

```
x + y
```

```
## Warning: Incompatible methods ("+.a", "+.b") for
## "+"
```

```
## [1] 4
## attr(,"class")
## [1] "a"
```

```
y + x
```

```
## Warning: Incompatible methods ("+.b", "+.a") for
## "+"
```

```
## [1] 4
## attr(,"class")
## [1] "b"
```

If both a and b have their own version of addition, then we need a way to resolve which version x+y and y+x should call. Here, the first operand takes precedence, so it determines which function is called—and you get a well-deserved warning for getting up to such shenanigans.

You *might* be able to come up with other rules on how to resolve such situations. For example, you could say that the most abstract method should be called, so if both x and y were of class b but only x was also of class a, then we should call +.b.

```
class(x) <- c("a","b")
class(y) <- "b"
```

Unfortunately, you cannot make such rules in the S3 system. It is possible with the S4 system, where you can dispatch generic functions based on multiple arguments, but that goes beyond the scope of this book. We will make sure to avoid situations where we have to add different classes of objects that define the same operators by constructing grammars appropriately.

Notice, though, that the combination of multiple classes and NextMethod still works as before. In the S3 system, we cannot define generic methods that dispatch based on multiple arguments, but we can use NextMethod to invoke several methods as we evaluate an operator. If we invoke +.a on object x, which now has classes a and b, the call to NextMethod in the implementation of that function will invoke +.b before *that* function invokes the numeric addition.

```
class(x) <- c("a", "b")
x + 2

## +.a
## +.bn

## [1] 3
## attr(,"class")
```

```
## [1] "a" "b"

x + y

## Warning: Incompatible methods ("+.a", "+.b") for
## "+"

## [1] 4
## attr(,"class")
## [1] "a" "b"
```

R has unary operators as well as binary operators. The negation operator, !, only exists in a unary operator, and you can specialize it for a given class by defining a function that takes a single argument.

```
`!.a` <- function(x) {
  cat("Not for a\n")
  NextMethod()
}
!x

## Not for a

## [1] FALSE
```

For the - and + operators, though, we use the same symbol for both unary and binary operators. Here, we have to use the same function for both since the only way we identify operator functions is through their name, and that would be the same for the unary and binary operators. The way we can determine whether the function is called as part of a unary or binary operator is to test whether the second argument is missing. If it is, we have a unary operator; otherwise, it is binary.

```
`+.a` <- function(e1, e2) {
  if (missing(e2)) {
    cat("Unary\n")
```

```
  } else {
    cat("Binary\n")
  }
  NextMethod()
}

class(x) <- "a"
+x

## Unary

## [1] 1
## attr(,"class")
## [1] "a"

2+x

## Binary

## [1] 3
## attr(,"class")
## [1] "a"
```

Group Generics

There is another way you can overload operators based on their operands'
class: group generics. Group generics, as the name suggests, group several
operators. They provide a way for us to define a single function that
handles all operators of a given type. For arithmetic and logical operators
(+, -, *, /, ^, %%, %/%, &, |, !, ==, !=, <, <=, >=, and >), the relevant group
generic is Ops.

    If we define Ops.c, then we define a function that will be called for all
of these operators when used on an element of class c.

```
Ops.c <- function(e1, e2) {
  cat(paste0("Ops.c (", .Generic, ")\n"))
  NextMethod()
}

z <- 2
class(z) <- "c"
z + 1

## Ops.c (+)

## [1] 3
## attr(,"class")
## [1] "c"

1 + z

## Ops.c (+)

## [1] 3
## attr(,"class")
## [1] "c"

z ^ 3

## Ops.c (^)

## [1] 8
## attr(,"class")
## [1] "c"
```

The "magical" variable .Generic contains the name of the operator that is called, and calling NextMethod will dispatch to the relevant next implementation of the operator.

If you implement both the Ops group generic *and* concrete implementations of some individual operator generics, then the latter takes precedence. If, for example, we have an object of class a *and* c, where we have defined addition for class a and have the group generic for c, the addition will invoke the +.a function. All other operators will invoke the Ops.c function.

```
class(z) <- c("a", "c")
1 + z

## Binary

## [1] 3
## attr(,"class")
## [1] "a" "c"

2 * z

## Ops.c (*)

## [1] 4
## attr(,"class")
## [1] "a" "c"
```

With Ops you have a method for catching all operators for which you do not explicitly write specialized generics.

Precedence and Evaluation Order

As soon as we start working with operators, their precedence becomes important. The syntax for normal function calls makes the evaluation order relatively clear—with nested function calls we have inner and outer functions in an expression. While that does not give us guarantees about which order parameters to a function will be evaluated in, we do know that arguments to a function will be evaluated before the function itself is evaluated.[1] With operators, the syntax does not tell us in which order functions will be called. To know that, we need to know the precedence rules.

Precedence rules tell us the order in which operator functions get called by ordering the operators from highest to lowest precedence and by specifying whether operators are evaluated from left to right or from right to left. In an expression such as this:

```
x + y * z
```

we know that the multiplication, y * z, is evaluated before the addition, so the expression is equivalent to the following:

```
x + (y * z)
```

because multiplication has higher precedence than addition. With operators at the same level of precedence, we might be less aware of the order, but here, the left-to-right or right-to-left order is also guaranteed by precedence rules. Of the operators you can overload, only the exponentiation operator evaluates right to left, while all the others evaluate left to right. Therefore, the following:

```
x ^ y ^ z
```

is equivalent to this:

```
x ^ (y ^ z)
```

And this:

```
x * y / z
```

will be evaluated as follows:

```
(x * y) / z
```

---

[1]I am not entirely honest here. R has lazy evaluation, so there is no guarantee that arguments to a function will be evaluated at all—but if they are, they will be evaluated before we return from the function they are arguments to, so conceptually we can think of them as being evaluated before we call the function.

The operators you can use are listed, from highest to lowest precedence, here:

Operator	Usual Meaning
[[[	Indexing
^	Exponentiation (evaluates right to left)
- +	Unary minus and plus
%any%	Special operators
* /	Multiply, divide
+ -	Binary add and subtract
< >	
<= >=	Ordering and comparison
== !=	
!	Negation
& &&	And
| ||	Or
:=	Assignment
-> ->>	Assignment
<- <<-	Assignment (right to left)
?	Help

Of these, you cannot overload the "arrow" assignment operators, only the : = operator.[2]

---

[2]The last operator in the table, the := assignment operator, is special in the list. It is not really an operator that is defined in R. You cannot use it as an assignment operator—for that, you should use <- or ->—but the R parser recognizes it as an infix operator, which means you can use it when you design a domain-specific language.

In the graph specifications language from the previous chapter, we could use addition and %=>% together because user-defined infix operators—those defined using percentages symbols—have higher precedence than +, so we would construct edges before we would add them to a graph. Using > and | together works for the same reason. Further, because addition (as well as logical or) is evaluated from left to right, the dag() object we created at the beginning of a graph specification would be added to the first edge, which would produce another graph that then would be added to the next edge and so forth. If the evaluation of addition were right to left, we would be adding edges to edges, instead of graphs to edges, which would complicate the implementation of the parser.

Code Blocks

A final syntactical component that can be useful when designing a domain-specific language is not an operator but the braces that construct a block of code. We cannot overload how these are interpreted, but we can certainly find a use for them when we create a new language. Before we can exploit them fully, we need to know both how to manipulate expressions and how to evaluate them in different contexts—which we cover in the next chapter and in Chapter 7—but as a quick example, consider creating an index operator that repeats a statement a number of times. We can define it this way:

```
`%times%` <- function(n, body) {
  body <- substitute(body)
  for (i in 1:n)
    eval(body, parent.frame())
}
```

The following takes the body argument and changes it to an expression so we can evaluate it repeatedly:

```
body <- substitute(body)
```

If we did not do this, we would evaluate it only the first time we accessed body—this is how R's lazy evaluation works—but using substitute we can get the verbatim expression out of the argument. To evaluate it, we then have to use the eval function, and to evaluate it in the context where we call the %times% operator, we need the calling frame, which we get using parent.frame (see Chapter 7 for more details on evaluation and environments).

The body argument to %times% can be a single statement, like this:

```
4 %times% cat("foo\n")
```

```
## foo
## foo
## foo
## foo
```

However, since braces ({}) are considered expressions as well, we can also use a sequence of statements as long as we wrap them in braces.

```
2 %times% {
  cat("foo\n")
  cat("bar\n")
}
```

```
## foo
## bar
## foo
## bar
```

Because we can use braces to pass blocks of code as arguments to functions, we can use these to create new control structures, like the `%times%` operator shown previously. To fully exploit this, we need to understand how we evaluate general expressions in R and how we control the environment in which we evaluate them. We also need to parse and manipulate expressions to analyze blocks of code and maybe modify them. We will leave further discussion of braces until we have covered those topics.

## CHAPTER 5

Parsing and Manipulating Expressions

A powerful feature of the R programming language is that it readily allows us to treat expressions in the language itself as data that we can examine and modify as part of a program—so-called meta-programming. From within a program we can take a piece of R code and computationally manipulate it before we evaluate it. We need to get hold of the code *before* it is evaluated, and there are several ways to do that. The simplest is to "quote" expressions, which leaves them as unevaluated expressions.

In this chapter, we will use the following libraries:

```
library(purrr)
library(rlang)
library(magrittr)
```

© Thomas Mailund 2018
T. Mailund, *Domain-Specific Languages in R*,
https://doi.org/10.1007/978-1-4842-3588-1_5

Quoting and Evaluating

If you write an expression such as the following, R will immediately try to evaluate it:

```
2 * x + y
```

It will look for the variables x and y in the current scope, and if it finds them, it will evaluate the expression; if it does not, it will report an error. By the time R has evaluated the expression, we have either a value or an error. If it is the former, the expression is essentially equivalent to the result of evaluating the expression (computation time notwithstanding). A literate expression as this one is not something we can get a hold on to in a program—we get either an error or the value the expression evaluates to. If we want to get hold of the actual expression, we need to "quote" it. If we wrap the expression in a call to the function quote, then we prevent the evaluation of the expression and instead get a data structure that represents the unevaluated expression.

```
quote(2 * x + y)
```

```
## 2 * x + y
```

The class of an expression is a "call."

```
expr <- quote(2 * x + y)
class(expr)
```

```
## [1] "call"
```

It is a call because infix operators are syntactic sugar for function calls, and all function call expressions will have this type. For "call" objects, we can get their components by indexing as we would a list. The first element will be the function name, and the remaining elements will be the arguments to the function call. For binary operators, of course, there will be two arguments.

For this expression, the function call is an addition:

```
expr[[1]]
```

```
## `+`
```

```
expr[[2]]
```

```
## 2 * x
```

```
expr[[3]]
```

```
## y
```

It is an addition because multiplication has higher precedence than addition, so the expression is equivalent to the following:

```
(2 * x) + y
```

This is because the multination is nested deeper in the expression than the addition. The multiplication can be accessed as the first argument to the addition call, so the second element in the object is as follows:

```
expr[[2]][[1]]
```

```
## `*`
```

```
expr[[2]][[2]]
```

```
## [1] 2
```

```
expr[[2]][[3]]
```

```
## x
```

To evaluate a quoted expression, we can use the function eval. The following expression:

```
eval(quote(2 * x + y))
```

is equivalent to writing the literate expression shown here:

```
2 * x + y
```

The eval function provides more flexibility in how an expression is evaluated since we can modify the scope of the evaluation, something we return to in more detail in Chapter 7.

Combining quoted expressions and functions introduces additional complications, at least if we want to handle the quoting within a function call. We can, however, pass quoted expressions as parameters to a function, as shown here:

```
f <- function(expr) expr[[1]]
f(quote(2 * x + y))
```

```
## `+`
```

However, it gets more complicated if we want to provide the literate expression to the function.

```
f(2 * x + y)
```

```
## Error in f(2 * x + y): object 'x' not found
```

In the function f, when we return expr[[1]], R will first attempt to evaluate the expression, but the expression depends on the variables x and y, which are undefined. Even if we define x and y, we still do not get a "call" object that we can manipulate. We get the result of evaluating the expression.

```
x <- 2
y <- 3
f(2 * x + y)
```

```
## [1] 7
```

Using quote inside the function doesn't help us. If we write quote(expr), we get the expression expr—a single symbol—as a result, not the argument we give to f.

```
f <- function(expr) {
  expr <- quote(expr)
  expr[[1]]
}
f(2 * x + y)
```

```
## Error in expr[[1]]: object of type 'symbol' is not subsettable
```

To get the actual argument as a quoted expression, we need to use the function substitute.

```
f <- function(expr) {
  expr <- substitute(expr)
  expr[[1]]
}
f(2 * x + y)
```

```
## `+`
```

Two things come together to make this work. First, function arguments in R are lazily evaluated, so the expr argument is never evaluated if we do not use it in an expression. So, even though x and y are not defined, we do not get any errors as long as we do not evaluate the argument to f. Second, substitute does not evaluate its argument, but it returns a quoted object where variables are replaced with the value they have in the current scope.[1] The argument to substitute does not have to be a single variable

---

[1]The substitute function will replace variables by the value they contain in the current scope or the value they have in an environment you provide as a second argument, *except* for variables in the global environment. Those variables are left alone. If you experiment with substitute, be aware that it behaves differently inside the scope of a function from how it behaves in the global scope.

name. It can be any expression that will be considered quoted after which variable substitution is done, and the return value will be the modified quoted expression.

```
f <- function(expr) {
  expr <- substitute(expr + expr)
  expr
}
f(2 * x + y)

## 2 * x + y + (2 * x + y)
```

Another complication appears if we attempt to evaluate a quoted expression inside a function. You might expect these two functions to be equivalent since eval(quote(expr)) should be the same as expr, but they are *not* equivalent.

```
f <- function(expr) {
  expr + expr
}
g <- function(expr) {
  x <- substitute(expr + expr)
  eval(x)
}
```

If we make sure that both x and y are defined, then the function f returns twice the value of the expression.

```
x <- 2; y <- 3
f(2 * x + y)

## [1] 14
```

Function g, on the other hand, raises an error because the type of x is incorrect.

```
g(2 * x + y)
```

```
## Error in 2 * x: non-numeric argument to binary operator
```

By default, the eval function will evaluate an expression in the current scope, which inside a function will be that function's evaluation environment. Inside g, we have defined x to be the expression we get from the call to substitute, so it is *this* x that is seen by eval. If you want eval to evaluate an expression in another scope, you need to give it an environment as a second argument. If you want it to evaluate the expression in the scope where the function is *called*, rather than inside the function scope itself, then you can get that using the parent.frame function.

```
g <- function(expr) {
  x <- substitute(expr + expr)
  eval(x, parent.frame())
}
g(2 * x + y)
```

```
## [1] 14
```

We will discuss environments, scopes, and how expressions are evaluated in more detail in Chapter 7. For the remainder of this chapter, we will focus on manipulating expressions and not on evaluating them.

Exploring Expressions

An expression is a recursive data structure, and you can explore it as such. We can define expressions in a grammar like this:

```
EXPRESSION ::= CONSTANT
            | NAME
            | PRIMITIVE
            | PAIRLIST
            | CALL EXPRESSION_LIST
EXPRESSION_LIST
          ::= EXPRESSION
            | EXPRESSION EXPRESSION_LIST
```

We will not expend the grammar of expressions further, but just agree that they will be any legal R expressions. All expressions are one of the five. The first four are terminals in the grammar, while call expressions are recursive; a call is constructed from a function and its arguments, and all these are other expressions.

We can explore expressions using recursive functions where the first three meta-variables, CONSTANT, NAME, and PRIMITIVE, are basis cases that do not contain other expressions, while PAIRLIST might and CALL will contain other expressions and must be handled in recursive calls.

Of the meta-variables, CONSTANT refers to any literal data such as numbers or strings, NAME refers to any variable name, PRIMTIVE refers to a function written in C as part of the implementation of R, PAIRLIST refers to formal arguments in function definitions (more on this below), and CALL refers to function calls. Function calls capture everything more complicated than the first four options. Since everything in R that does anything is considered a function call, including such statements as function definitions and control structures, these are captured in the CALL case. As we saw earlier, calls are list-like and always have at least one element. The first element is the function that is called, and the remaining components are the arguments to that function.

To recursively explore an expression, we can write functions that test the four cases. Constants are recognized by the is.atomic function, names by the is.name function, primitives by the is.primitive function, pair lists by the is.pairlist, and calls by the is.call function. A function for printing out an expression's structure can look like this:

```
print_expression <- function(expr, indent = "") {
  if (is.atomic(expr)) {
    if (inherits(expr, "srcref")) {
      expr <- paste0("srcref = ", expr)
    }
    cat(indent, " - ", expr, "\n")

  } else if (is.name(expr)) {
    if (expr == "") {
      expr <- "MISSING"
    }
    cat(indent, " - ", expr, "\n")

  } else if (is.primitive(expr)) {
    cat(indent, " - ", expr, "\n")

  } else if (is.pairlist(expr)) {
    cat(indent, " - ", "[\n")
    new_indent <- paste0(indent, "        ")
    vars <- names(expr)
    for (i in seq_along(expr)) {
      cat(indent, "    ", vars[i], " ->\n")
      print_expression((expr[[i]]), new_indent)
    }
    cat(indent, "     ]\n")
```

```
  } else {
    print_expression((expr[[1]]), indent)
    new_indent <- paste0("  ", indent)
    for (i in 2:length(expr)) {
      print_expression(expr[[i]], new_indent)
    }
  }
}
```

Here, we do not explicitly test for the type of calls; if the expression is not one of the first four cases, it must be the fifth. There are two special cases we handle in this printing expression—source references for function definitions and missing expressions in pair lists. We discuss these next.

We can see the function in action by calling it on the expression we explored earlier.

```
print_expression(quote(2 * x + y))
```

```
##   - +
##     - *
##       - 2
##       - x
##     - y
```

The pretty-printed expression shows the structure we explored explicitly in the previous section.

Declaring a function is considered a function call—a call to the function function.

```
print_expression(quote(function(x) x))
```

```
##   - function
##     - [
##         x  ->
##           - MISSING
```

```
##            ]
##       -   x
##       -   srcref = function(x) x
```

For a function definition, we have a call object where the first argument is `function`, the second argument is the pair list that defines the function parameters, and the third element is the function body—another expression. There is also a fourth element called `srcdef`, an atomic vector that captures the actual code used to define the function. In the printing function, we just print the text representation of the source definition, which we get by pasting the expression.

The argument list of a function we declare is where the pair list data structure is used. We can get the names of the formal parameters using the `names` function and the default arguments by indexing into the pair list. Parameters without default arguments are a special case here, and the expression they contain is an empty string. In the printing function, we make this explicit by changing the empty string to the string `MISSING`. If we have default arguments, then those are represented as expressions we can explore recursively.

print_expression(quote(function(x = 2 * 2 + 4) x))

```
##     -   function
##        -   [
##            x   ->
##                 -   +
##                  -   *
##                    -   2
##                    -   2
##                   -   4
##            ]
##        -   x
##        -   srcref = function(x = 2 * 2 + 4) x
```

```
print_expression(quote(function(x, y = 2 * x) x + y))
```

```
##    -  function
##      -  [
##          x  ->
##                -  MISSING
##          y  ->
##              -  *
##                  -  2
##                  -  x
##          ]
##      -  +
##        -  x
##        -  y
##      -  srcref = function(x, y = 2 * x) x + y
```

The usual case for function calls is that the first element in the "call" list is a symbol that refers to a function, and any expression that returns a function can be used as a function in R. This means the first element of calls can be any expression. For example, if we define a function and call it right after, the first element of the call object will be the function definition.

```
expr <- quote((function(x) x)(2))
print_expression(expr)
```

```
##    -  (
##      -  function
##        -  [
##            x  ->
##                -  MISSING
##            ]
##        -  x
##      -  srcref = function(x) x
##    -  2
```

```
expr[[1]]
```

```
## (function(x) x)
```

```
expr[[2]]
```

```
## [1] 2
```

As an example of doing something non-trivial with expressions, we can write a function that collects all unbound variables in an expression. If we recurse through an expression, we can collect all the bound and unbound symbols. To get the unbound variables, we can keep track of those that are bound and not collect those. Ignoring, at first, those variables that might be bound outside of the expression itself—in the scope where we will call the function—the variables that are bound are those that are named in a function definition. We can identify those from the pair list that is the second argument to calls to function. When recursing over expressions, we capture those and pass them on down the recursion. Aside from that, we simply collect the symbols. In the following implementation, I use the linked lists we saw earlier to collect the symbols, and I translate the symbols into characters as I collect them. I do this because I can use the character representation of symbols to check whether a symbol exists in an environment later. I use the cons function to collect symbols in a linked list.

```
cons <- function(car, cdr) list(car = car, cdr = cdr)
collect_symbols_rec <- function(expr, lst, bound) {
  if (is.symbol(expr) && expr != "") {
    if (as.character(expr) %in% bound) lst
    else cons(as.character(expr), lst)

  } else if (is.pairlist(expr)) {
    for (i in seq_along(expr)) {
      lst <- collect_symbols_rec(expr[[i]], lst, bound)
    }
    lst
```

```r
  } else if (is.call(expr)) {
    if (expr[[1]] == as.symbol("function"))
      bound <- c(names(expr[[2]]), bound)

    for (i in 1:length(expr)) {
      lst <- collect_symbols_rec(expr[[i]], lst, bound)
    }
    lst

  } else {
    lst
  }
}
```

For processing the lists, it is easier to work with list than with linked-lists objects, so we need the lst_to_list function from earlier as well.

```r
lst_length <- function(lst) {
  len <- 0
  while (!is.null(lst)) {
    lst <- lst$cdr
    len <- len + 1
  }
  len
}
lst_to_list <- function(lst) {
  v <- vector(mode = "list", length = lst_length(lst))
  index <- 1
  while (!is.null(lst)) {
    v[[index]] <- lst$car
    lst <- lst$cdr
    index <- index + 1
  }
  v
}
```

We explicitly avoid the empty symbol when we collect symbols. The empty symbol is the symbol we get when we recurse on a pair list for a function parameter without a default value. We do not consider this a variable, bound or otherwise. The way we handle symbols is straightforward. For pair lists, we collect the parameters that will be bound and recurse through the default arguments to collect any unbound variables there. As for calls, we handle the function definitions by extending the list of bound variables and then recursing. For anything else—which in practice means for any atomic value—we just return the list we called the function with. There are no unbound variables in constant values after all.

The recursive function works on a quoted expression and collects all symbols that are not bound within the expression itself. We wrap it in a function that does the quoting of the expression, call the recursive function, and then remove the symbols that are defined in the calling scope (the parent.frame).

```
collect_symbols <- function(expr) {
  expr <- substitute(expr)
  bound <- c()
  lst <- collect_symbols_rec(expr, NULL, bound)
  lst %>% lst_to_list() %>% unique() %>%
         purrr::discard(exists, parent.frame()) %>%
         unlist()
}
```

Here, I use the discard function from the purrr package to remove all elements that satisfy a predicate. For the predicate, I use the function exists with a second argument that is the calling environment, parent. frame. This gets rid of symbols that are defined in the scope where we call collect_symbols, including globally defined functions such as *, +, and function.

I pipe the final result through `unlist` to translate the `list` into a character vector. This is only for pretty-printing reasons. It gives nicer output when printed in the console. For programming, you can work with `list`s as well as with vectors.

If we get rid of variables x and y that we defined earlier, the expression 2 * x + y + z should have three unbound variables, x, y, and z. This is indeed what we find:

```
rm(x) ; rm(y)
collect_symbols(2 * x + y + z)
```

```
## [1] "z" "y" "x"
```

If we define one of the variables, for example, z, then it is no longer unbound.

```
z <- 3
collect_symbols(2 * x + y + z)
```

```
## [1] "y" "x"
```

Function definitions also bind variables, so those are not collected.

```
collect_symbols(function(x) 2 * x + y + z)
```

```
## [1] "y"
```

```
collect_symbols(function(x) function(y) f(2 * x + y))
```

```
## NULL
```

Default values can contain unbound variables; we collect those values.

```
collect_symbols(function(x, y = 2 * w) 2 * x + y)
```

```
## [1] "w"
```

We are not entirely done learning about how to explore expressions yet. The actual recursive exploration of expressions is simple, as shown previously. But often, it must be combined with an evaluation of expressions. And often, this evaluation does not follow the usual rules for how expressions are evaluated because we have to evaluate some expressions while we keep others quoted. When we start manipulating how expressions are evaluated, we call it *non-standard evaluation*, which is the topic of Chapter 7. Here, however, I want to give you a taste of what it involves.

If we write a simple function such as this:

```r
f <- function(expr) collect_symbols(expr)
```

we might expect it to give us the unbound variables in an expression, but it returns an empty list, as shown here:

```r
f(2 + y * w)
```

```
## NULL
```

This is because of the combination of the two issues we will have when we try to program the functions of the so-called non-standard evaluation. First, when we use substitute in the collect_symbols function, we get the literal expression that substitute was called with. The argument we give to substitute in f is expr. The expression that f itself is called with does not get passed along. Second, the environment in which we test for a bound variable inside collect_symbols is the calling environment. When we call the function from f, the calling environment is the body of f. In this environment, the variable expr is defined—it is the formal argument of the function—so it will be considered bound.

We will explore environments and how to program with non-standard evaluation in some detail later, but the general solution to these problems is to avoid using non-standard evaluation in functions you plan to call from other functions. It is a powerful technique for writing a domain-specific language, but keep it to the interface of the language and not the internal functions. For collect_symbols, we can get around the problem by writing another function that takes as arguments a quoted expression and an environment we should look for variables in. We can then call this function from collect_symbols when we want a non-standard evaluation and call the other function directly if we want to use it from other functions.

```
collect_symbols_ <- function(expr, env) {
  bound <- c()
  lst <- collect_symbols_rec(expr, NULL, bound)
  lst %>% lst_to_list() %>% unique() %>%
    purrr::discard(exists, env) %>%
    unlist()
}
collect_symbols <- function(expr) {
  collect_symbols_(substitute(expr), parent.frame())
}
```

Manipulating Expressions

We can do more than simply inspect expressions. We can also modify them or create new ones from within programs. You cannot modify the two primitive expressions, constants and symbols. They are simply data. We can, however, modify calls and pair lists, although the second is not something we would usually do. We work with pair lists when we create new functions, but usually we either create new pair lists to set the formal arguments of a function or take the arguments from another function; we rarely modify existing pair lists. In any case, both pair lists and calls can be assigned to by indexing into their components.

To get it out of the way with, the following is an example where we modify a pair list. We can construct the expression for defining a function like this:

```
f <- quote(function(x) 2 * x)
f
```

```
## function(x) 2 * x
```

This is an expression of the type "call"—it is a call to the function function that defines functions (try saying that fast)—and its second argument is the pair list that defines its arguments.

```
f[[2]]
```

```
## $x
```

If we assign to the elements in this pair list, we provide default arguments to the function. The values we assign must be quoted expressions.

```
f[[2]][[1]] <- quote(2 * y)
f
```

```
## function(x = 2 * y) 2 * x
```

To change the names of function arguments, we must change the names of the pair list components. We can do this using the names<- function.

```
names(f[[2]]) <- c("a")
f[[3]] <- quote(2 * a)
f
```

```
## function(a = 2 * y) 2 * a
```

In this example, we also saw how we could modify the function body through its third component.

Through this example, we have already seen all we need to know about how to modify call expressions. What we were modifying was simply a particular case of a call—the call to `function`. Any other call can be changed the same way.

```
expr <- quote(2 * x + y)
expr
```

```
## 2 * x + y
```

```
expr[[1]] <- as.symbol("/")
expr
```

```
## 2 * x/y
```

```
expr[[2]][[1]] <- as.symbol("+")
expr
```

```
## (2 + x)/y
```

We can construct new call objects using the `call` function. As its first argument, this function takes the function to call. This can be a symbol or a string and will automatically be quoted. After that, you can give it a variable number of arguments that will be evaluated before they are put into the constructed expression.

```
call("+", quote(2 * x), quote(y))
```

```
## 2 * x + y
```

```
call("+", call("*", 2, quote(x)), quote(y))
```

```
## 2 * x + y
```

If you are creating a call to a function with named arguments, rather than an operator, you can provide those to the `call` function as well.

```
call("f", a = quote(2 * x), b = quote(y))
```

```
## f(a = 2 * x, b = y)
```

It is essential that you quote the arguments if you do not want them evaluated. The call function will not do it for you.

```
z <- 2
call("+", 2 * z, quote(y))
```

```
## 4 + y
```

In the rlang package you have two additional functions for creating calls. The function lang works as the call function except that you can specify a namespace in which the called function should be found. The new_language function lets you provide the call arguments as an explicit pair list.

```
library(rlang)
lang("+", quote(2 * x), quote(y))
```

```
## 2 * x + y
```

```
new_language(as.symbol("+"), pairlist(quote(2 * x), quote(y)))
```

```
## 2 * x + y
```

The rlang package is worth exploring if you plan to do much meta-programming in R. It provides several functions for manipulating and creating expressions and functions and for managing environments. We will explore the package more in Chapter 8.

There is one extra complication if the call you are making is to function. This function needs a pair list as its second argument, so you will have to make such an object. If you want to create a function without default parameters, you need to make a list with "missing" elements at named positions. The way to make a missing argument is by calling substitute without arguments, so a function that creates a list of function parameters without default arguments can look like this:

```
make_args_list <- function(args) {
  res <- replicate(length(args), substitute())
  names(res) <- args
  as.pairlist(res)
}
```

We can use it to construct a call to function like this:

```
f <- call("function",
          make_args_list(c("x", "y")),
          quote(2 * x + y))
f
```

```
## function(x, y) 2 * x + y
```

Remember, however, that this is an expression for creating a function; it is not the function itself, and it does not behave like a function.

```
f(2, 3)
```

```
## Error in f(2, 3): could not find function "f"
```

The error message here looks a bit odd. R is not complaining that f is not a function but that the function f cannot be found. This is because R will look for functions when you use a symbol for a function call and will not confuse the value f with the function f. Here, we only have a value-version of f. Anyway, to get the actual function, we need to evaluate the call.

```
f <- eval(f)
f
```

```
## function (x, y)
## 2 * x + y
```

```
f(2, 3)
```

```
## [1] 7
```

A more direct way of creating a function is by using the `new_function` function from the `rlang` package.

```
f <- new_function(make_args_list(c("x", "y")),
                  quote(2 * x + y))
f
```

```
## function (x, y)
## 2 * x + y
```

```
f(2, 3)
```

```
## [1] 7
```

As a final example, we can combine the expression creating methods we have seen with the expression exploration functions from the previous section to translate expressions with unbound variables into functions. We can collect all unbound variables in an expression using the `collect_symbols_` function from earlier and then use `new_function` to create the function.

```
expr_to_function <- function(expr) {
  expr <- substitute(expr)
  unbound <- collect_symbols_(expr, caller_env())
  new_function(make_args_list(unbound), expr, caller_env())
}
```

Here, I have used another function from `rlang`, `caller_env`. This function does the same as the `parent.frame` we have used earlier but with a more informative name. I recommend using `caller_env` over `parent. frame` for that reason.

We provide more arguments in this call to new_function than in the previous example where we used it. There, we provided only two arguments, the parameters of the function and its body. Here, we also provide its environment. This will be the function's enclosing environment. It is here that the function will find the value of variables that are not local to the function itself or parameters to the function. Since we consider variables found in the caller environment as bound, we have to make sure that the function we create can also find them, so we put the function in the same environment. If this explanation is unclear to you, then return to this example after you have read Chapter 7 where we go into environments in much more detail. It should, ideally, be clearer then.

expr_to_function does exactly what we intended it to do. It creates a function from an expression, whose arguments are the unbound variables.

```
f <- expr_to_function(2 * x + y)
f

## function (y, x)
## 2 * x + y

f(x = 2, y = 3)

## [1] 7

g <- expr_to_function(function(x) 2 * x + y)
g

## function (y)
## function(x) 2 * x + y

g(y = 3)(x = 2)

## [1] 7
```

The order of the variables in the function will depend on the order in which they appear in the expression and in whatever order the unique function will leave them in. Therefore, calling the resulting function is best done with named arguments.

100

CHAPTER 6

Lambda Expressions

With the techniques we have seen so far, we are now able to implement some useful domain-specific languages. In this chapter, we examine a toy example, lambda expressions. It is perhaps not something we would use in real-world code, as it simply gives an alternative syntax to anonymous functions, which are already supported in R. However, it is an excellent example of code that is potentially useful and gives us a chance to experiment with syntax.

We will use the `rlang` package.

```
library(rlang)
```

Anonymous functions

Lambda expressions are Anonymous functions, in other words, functions we have not named. We already have anonymous functions in R. This is the default kind of functions since a function is anonymous until we assign it to a variable. If we do not want to save a function in a variable to get access to it later, we can just use the `function` expression to create it where we need it. For example, to map over a vector of numbers, we could write the following:

```
sapply(1:4, function(x) x**2)
```

```
## [1]  1  4  9 16
```

© Thomas Mailund 2018
T. Mailund, *Domain-Specific Languages in R*,
https://doi.org/10.1007/978-1-4842-3588-1_6

This is a toy example since vector expressions are preferable in situations like the following, but it illustrates the point.

```
(1:4)**2
```

```
## [1]  1  4  9 16
```

Using `function` expressions is verbose, so we might want to construct an alternative syntax for anonymous functions. We can then use it as an exercise in constructing a domain-specific language. Our goal is to change the previous `sapply` syntax into this syntax:

```
sapply(1:4, x := x**2)
```

We use the `:=` assignment operator for two reasons. One, we can overload it, something we cannot do with `->` or `<-`. Two, it has the lowest precedence of the operators, so the operator we create will be called with the left- and right-hand sides before these are evaluated.

To implement this syntax, we need to make the left-hand side of assignments into function headers, which means pair lists of arguments. We also need to make the right-hand side into a function body we can evaluate in the environment where we define the lambda expression. The good news is that this only involves techniques we have already seen. We can write a function for turning a list of arguments into a pair list that we can use to define the formal arguments of a function like this:

```
make_args_list <- function(args) {
  res <- replicate(length(args), substitute())
  names(res) <- args
  as.pairlist(res)
}
```

For the assignment operator, we need to use `substitute` to avoid evaluating the two arguments. We then use `make_args_list` to turn the left-hand side into formal arguments, but we keep the right-hand side expression as it is. After that, we turn the combination into a function using new_function from the rlang package. Since we want to evaluate the new function in the scope where we define the lambda expression, we use caller_env to get this environment and provide it to new_function. The entire implementation is as simple as this:

```
`:=` <- function(header, body) {
  header <- substitute(header)
  body <- substitute(body)
  args <- make_args_list(as.character(header))
  new_function(args, body, caller_env())
}
```

Now, we can use the new syntax as syntactic sugar for anonymous functions.

```
sapply(1:4, x := x**2)
```

```
## [1]  1  4  9 16
```

What about lambda expressions with more than one argument? We might want syntax similar to this:

```
mapply(x,y := x*y, x = 1:6, y = 1:2)
```

However, this is not possible since we cannot override how R interprets commas. If we want to group some parameters, we need to put them in a function call. We can do something like this:

```
mapply(.(x,y) := x*y, x = 1:6, y = 1:2)
```

```
## [1]  1  4  3  8  5 12
```

What happens here is that the make_args_list translates all the components of the left-hand expression into function parameters. A function call object is just like any other expression list, so in this particular

example, we create a function with three arguments, ., x, and y. Since
. is not used inside the function body, it does not matter that we do not
provide it when the function is called. However, if we reuse one of the
parameter names as the function name in the call, this happens:

```
mapply(x(x,y) := x*y, x = 1:6, y = 1:2)
```

```
## Error in (function (x, x, y) : argument 1 matches multiple
formal arguments
```

We can get rid of the function name in calls by removing the first
element in the list.

```
`:=` <- function(header, body) {
  header <- substitute(header)
  if (is.call(header)) header <- header[-1]
  body <- substitute(body)
  args <- make_args_list(as.character(header))
  new_function(args, body, caller_env())
}
```

Now the earlier example will work.

```
mapply(x(x,y) := x*y, x = 1:6, y = 1:2)
```

```
## [1]  1  4  3  8  5 12
```

Experiments with Alternatives to the Syntax

Using an assignment operator to define a function in this way might not be
the most obvious syntax you could choose, but we have plenty of options
for playing around with alternatives.

We could start with the functionality that we have implemented as a single function. There is no reason to have a special syntax if all we need is a single function, so instead, we could implement lambda expressions like this:

```
lambda <- function(...) {
  spec <- eval(substitute(alist(...)))
  n <- length(spec)
  args <- make_args_list(spec[-n])
  body <- spec[[n]]
  new_function(args, body, caller_env())
}
```

The idea here is that the lambda function will take a list of arguments where the last element in the list is the function body and the preceding are the parameters of the lambda expression.

```
sapply(1:4, lambda(x, 4 * x**2))
```

```
## [1]  4 16 36 64
```

```
mapply(lambda(x, y, y*x), x = 1:4, y = 4:7)
```

```
## [1]  4 10 18 28
```

The eval(substitute(alist(...))) expression might look a little odd if you are not used to it. What we do is take the variable number of arguments, captured by the three dots argument, and create an expression that turns those into a list. The function alist, unlike list, will not evaluate the expressions but keep the arguments as they are, which is what we want in this case. The substitute expression only creates the expression, so we need to evaluate it with eval to get the actual list. Once we have the list, we make the first arguments into function parameters and the last into the body of the lambda expression and create the function.

In production code, we should add some checks to make sure that the lambda expression parameters are symbols and not general expressions. However, the full functionality for lambda expressions is present in the function we have just written.

Of course, the `lambda` function does not behave like a normal function. The non-standard evaluation (NSE) we apply to make a function out of the arguments to `lambda` is very different from how functions normally behave, where the arguments we provide are considered values rather than symbols and expressions. To make it clear from the syntax that something different is happening, you could change the syntax. For example, we could go for square brackets instead of parentheses. We can implement a version that uses those like this:

```
lambda <- structure(NA, class = "lambda")
`[.lambda` <- function(x, ...) {
  spec <- eval(substitute(alist(...)))
  n <- length(spec)
  args <- make_args_list(spec[-n])
  body <- spec[[n]]
  new_function(args, body, caller_env())
}
```

We use it like this:

```
sapply(1:4, lambda[x, 4 * x**2])
```

```
## [1]  4 16 36 64
```

```
mapply(lambda[x, y, y*x], x = 1:4, y = 4:7)
```

```
## [1]  4 10 18 28
```

The approach here is to make `lambda` an object with a class we can use for defining a special case of the subscript operator. The sole purpose of `lambda` is to dispatch the subscript function to the right specialization, and that specialization of the subscript operator is the one that creates the new function. The only difference is that it takes an extra first argument, which is the `lambda` object. We do not use it for anything, so we just ignore it.

Don't Do This at Home

Implementing syntactic sugar for lambda expressions as we have done only saves us minimal typing compared to using function expressions. Those familiar with function expressions should know that this will potentially do more harm than good, but it might not be the case with our home-made syntax for them. Consequently, I do not recommend that you construct a new syntax for language constructions that are already implemented in R. We implemented the lambda expressions here to illustrate how we can construct new syntax with very little code.

CHAPTER 7

Environments and Expressions

We have already used environments in a couple of examples to evaluate expressions in a different context than where we usually evaluate them, which is known as *non-standard evaluation*. Many domain-specific languages that we could implement in R will need some variety of non-standard evaluation, but getting the evaluation to occur in the right context can be problematic. The rules for how expressions are evaluated are simple, while evaluation contexts, which are chains of environments, can be complicated.

We will use the `rlang` package.

```
library(rlang)
```

Scopes and Environments

R evaluates an expression in a *scope* that determines which value any given variable refers to. In the standard evaluation, R uses what is known as *lexical scope*. This essentially means that variables in an expression are referring to the variables defined in the blocks around the expression. If you write an expression at the outermost level of an R script, or in the *global environment*, then variable names in the expression refer to global variables. An expression inside a function, on the other hand, is evaluated in the scope of a function execution, which means that variable symbols refer to local

© Thomas Mailund 2018
T. Mailund, *Domain-Specific Languages in R*,
https://doi.org/10.1007/978-1-4842-3588-1_7

variables or function parameters if they are defined; only if they are not defined do they refer to global variables. A function defined inside another function will have a *nested scope*—variables in an expression there will first be searched for in the innermost function, then the surrounding function, and only if they are not found either place, in the global environment.

Consider this abstract example:

```
x <- 1
f <- function(y) {
  z <- 3
  function() x + y + z
}
g <- f(2)
h <- f(3)
g()

## [1] 6

h()

## [1] 7
```

In the example, we define four variables in the global environment, x, f, g, and h. In the function f we have one formal parameter, y, and one local variable, z. Whenever we call f, a scope where y exists is created, and the first statement in the function call adds z to this scope. The function returns another function, a closure, that contains an expression that refers to variables x, y, and z. If we call this function, which we do when we call functions g and h that are the results of two separate calls to f, we will evaluate this expression. When R evaluates the expression, it needs to find the three variables. They are neither formal arguments nor local variables in the functions we call (g and h), but since the functions were created inside calls to f, they can see y and z in their surrounding scope. Both can find x in the global environment. Since g and h are the results of separate calls to f, the surrounding scope of calls to them are *different* instances of local scopes of f.

110

Scopes are implemented through *environments*, and even though the rules that guide environments and evaluation are straightforward, you have to be careful if you start manipulating them. You can think of environments as tables that map variables to values. Also, all environments have a parent environment, an enclosing scope, that R will search in if a variable is not found when it searches the first environment. Environments thus have a tree structure that usually follows nested scopes, and that ends in a root in the *empty environment*. Packages you load are put on top of this environment, and on top of all loaded environments, we have the *global environment*—which is why you can find variables defined in packages if you search in the global environment.

Strictly speaking, there are a few other details on how packages and environments interact that I do not include in this view on environments, but they are not important for the discussion here. If you are interested, you can find these details in my other book, *Meta-programming in R* (Mailund, 2017c). For this book, we will simply assume that everything we define at the global level or any package is found in the global environment and consider this the root of the environment tree.

When we define new functions, we do not create new environments, but we do associate the functions with one—the environment in which we define the function. When we defined function f in the previous example, it got associated with the global environment, because that is where we defined it. We can get the environment a function is associated with using the environment function.

```
environment(f)
```

```
## <environment: R_GlobalEnv>
```

Since f is defined at the global level, its environment is the global environment. When we make a function call, we create a new environment called the *execution environment*. This environment is where we store parameters and local variables. The environment associated with the function will be the *parent environment* for this execution environment.

When we call function f, we thus create an environment where we get a mapping from y and z to their values and with a parent environment that is the global environment, in which we can find the variable x. Inside the call to f, we create a new (anonymous) function and return it. This function will also have an environment associated with it, but this time it is the local environment we created when we called f. Thus, the environments associated with g and h are two different environments as they are the result of two different calls to f.

environment(g)

```
## <environment: 0x7fdd80b25f08>
```

environment(h)

```
## <environment: 0x7fdd80a7f468>
```

Functions defined inside other functions thus carry along with them the environments that were created when the surrounding function was called, and if we return them from the surrounding function, they still carry this enclosing scope along with them. Since they remember the local variable and parameters from the enclosing scope, we call such functions *closures.*

In Figure 7-1, I have drawn a simplified graph that shows which environments exist and how they are wired together in the example at the point where we call function g. I show environments with a gray background, variables as circles with pointers to the values the variables refer to, and functions as the three components that define a function: the formal parameters, the function body, and the enclosing scope—the environment associated with the function.

The enclosing environment for function f is the global environment, while the enclosing scopes of g and h are the two different instances of calls to f. These instance or execution environments have the global environment as their parents since that is the enclosing scope of f. Because they are two different instances of f, the variables in them can point to different values, as we see for the variable y. For function g, y points to 3, while for function h, y points to 2. In a call to function g, we create a local environment for the function call—shown at the bottom right in the figure. We do not have any local variables in g, so this environment does not contain any variables, but it has a parent that is the f instance where we created g.

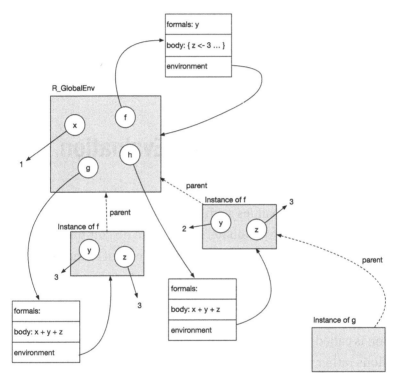

Figure 7-1. *Environment graph when calling g*

When we evaluate the expression x + y + z inside the call to g, we need to map variables to values. The search will start in the local environment and then progress up the parent links until it finds a matching variable. For variables y and z we find values in the parent of the g call, the instance of the f call that created g. For x we find the value in the grandparent, the global environment.

The rule for evaluating expressions is always the same: we look for variables by searching in environments, starting with the immediate environment where we evaluate the expression and search along the chain of parent environments. We check each environment in the chain in turn until we find the variable we are looking for. We get the standard evaluation rules of lexical scoping because functions get associated with the environment where they are created and since this environment is set as the parent environment of execution environments. The only trick to understanding how expressions are evaluated in R is to understand which environments are used. For the body of functions, it is as simple as I have just explained, but for function parameters, there are a few more rules to consider.

Default Parameters, Lazy Evaluation, and Promises

When you pass primitive values such as numbers to a function parameter, there is nothing we need to evaluate, so there are no complications. This is why we didn't have to worry about the environment of the arguments in the previous example. If we pass expressions along as parameters, however, we need to know how they should be evaluated.

Most of the time, R behaves as if expressions are evaluated before a function is called, but this not what happens. If we passed values to functions rather than expressions, we would not be able to get the expressions using `substitute` as we have done in previous chapters. When we call a function in R, the parameters will refer to unevaluated expressions; such expressions are known as *promises*. Promises are evaluated the first time we use a parameter variable but not before—an approach to parameter evaluation known as *lazy evaluation*. If we never refer to an argument, the corresponding expression will never be evaluated, so we can write code such as this without raising exceptions:

```
f <- function(x, y) x
f(2, stop("error!"))
```

```
## [1] 2
```

We never refer to the parameter y inside the body of f, so we never evaluate it. Consequently, we never call stop to raise the error.

So, since parameters can contain expressions, we need a rule for how to evaluate them. Here, there is a difference between default parameters, defined when the function is created, and parameters provided when the function is called. The former is evaluated in the local scopes of function calls, while the latter is evaluated in the environment where the function is called.

Consider this function:

```
f <- function(y, z = 2 * y) y + z
```

The function takes two parameters, y and z.

```
f(2, 1)
```

```
## [1] 3
```

But if we only provide y, then z will be set to 2 * y.

```
f(2)
```

```
## [1] 6
```

When we evaluate the promise that z points to when the function is called—we do this in the expression where we use the variable—the promise expression is evaluated. This means that R needs to find the variable y. If we try to evaluate the expression 2 * y in the scope where the function is defined—the global environment—then we would get an error, as there is no y variable defined there. The semantics of default parameters *could* be such that we evaluated them in the scope where we define a function. If so, we wouldn't be able to make default parameters depend on other parameters, which is what we want here—we want z to depend on y if we do not explicitly provide a value to it. The actual semantics is that the promise is evaluated in the function-call environment. When we call f, before we evaluate the y + z expression, the situation is therefore as shown in Figure 7-2. Here, I have drawn the promise for z as the expression passed along as the function argument together with the environment in which it should be evaluated.

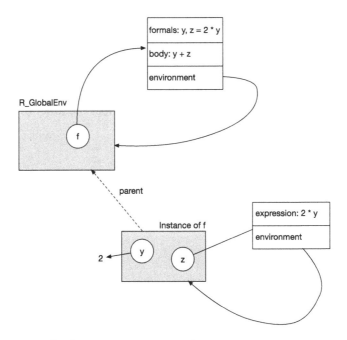

Figure 7-2. *Default parameter promise*

When we call f with a parameter that is an expression, we do *not* want to evaluate this expression in function-call scope. Consider this:

```
y <- 2
f(2 * y)

## [1] 12
```

The intent is to call f with 2 * y, which should be 4 since y is 2. If we tried to evaluate it inside the function call, however, we would have a circular dependency. Inside the function call, y is a variable, and if it points to 2 * y, we cannot evaluate the expression without knowing what y is, which we cannot know until we have evaluated the expression, which we cannot because we do not know what y is....

When we call a function with an expression as an argument, the corresponding promise will be evaluated in the environment where we call the function, so before we evaluate y + z inside f, the situation is as shown in Figure 7-3. Inside the environment of the function call, both y and z refer to promises, but these promises are associated with different environments. To evaluate the expression y + z, we need to evaluate both promises. To get the value for y, we need to evaluate 2 * y *in the global scope*, which gives us 4, and to get the value for z we need to evaluate 2 * y *in the local scope*, which gives us 8.

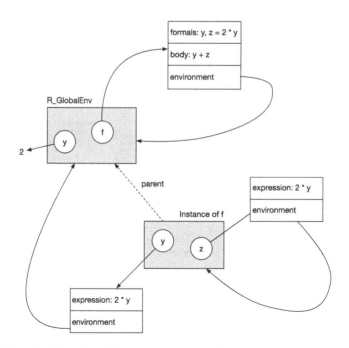

Figure 7-3. *Calling f with an expression for y*

At the risk of taking the example a step too far, let's consider the situation where we call f from another function.

```
g <- function(x) f(2 * x)
g(2 * y)
```

```
## [1] 24
```

Before we evaluate the expression y + z inside function f, the state of the environment graph is as shown in Figure 7-4. It takes a little effort to see what happens when we want to evaluate y + z, but doing this exercise will go a long way toward understanding environments.

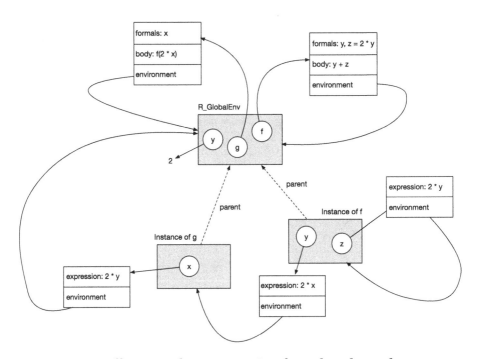

Figure 7-4. *Calling g with an expression for x that depend on variable y*

We have not yet evaluated the promises y and z. Since z depends on y, we need to evaluate y first. To do this, we need to evaluate the expression 2 * x in the scope of the call to g. Here, we need to evaluate x, which is another promise: the expression 2 * y that should be evaluated in the global scope (where y refers to a different variable than the local variable inside the f instance). In the global scope, y refers to the value 2, so we can evaluate 2 * y directly and get the value 4. This value then replaces the promise in the scope of the call to g. Once we have evaluated a promise, the variable refers to the value and no longer the expression. This now means that we can evaluate 2 * x in the scope of the g call to get 8. So now y in the call to f refers to 8. This means we can evaluate 2 * y to get 16, which we assign the variable z. Finally, we can evaluate y + z to get $8 + 16 = 24$.

To summarize this section, parameters we pass to functions, if they are not primitive values, are considered expressions that must be evaluated at some point. Associated with the expressions, we have a scope in which to evaluate the them. There is one more caveat, though, which I hinted to in the previous example: parameters are considered expressions only until the first time we evaluate them. After that, they are the result of this evaluation.

There are pros and cons with these semantics—though predominantly cons. We can avoid computing values we do not need because promises are not evaluated until we refer to the variable that holds them. Moreover, we can make default parameters that depend on some computation inside a function call as long as we do those computations before we use the variable that needs them. For example, we can define a default parameter in terms of a variable we set inside a function.

```r
h <- function(x, y = 2 * w) {
  w <- 2
  x + y
}
h(1)

## [1] 5
```

However, this will fail if we refer to the promise that needs the variable before we compute it.

```
h <- function(x, y = 2 * w) {
  res <- x + y
  w <- 2
  res
}
h(1)
```

```
## Error in h(1): object 'w' not found
```

We have to be careful if a promise depends on a variable that we update during a computation. Note that a promise is evaluated only once; after the evaluation, the variable that used to hold it now holds the result of the evaluation and no longer the promise expression. If we change variables that occurred in the promise, we do not update the value that the variable now holds.

```
h <- function(x, y = 2 * w) {
  w <- 1
  res <- x + y
  w <- 2
  res
}
h(1)
```

```
## [1] 3
```

The promises held by default parameters do not usually cause problems. It is simple to follow which local variables will change in the function and at what point the promise will be evaluated. Lazy evaluation of arguments, however, is a common source of problems when combined with closures. Consider this function:

```
make_adder <- function(n) function(m) n + m
```

This returns a closure that will add n to its argument, m. We can use it like this:

```
add_1 <- make_adder(1)
add_2 <- make_adder(2)
add_1(1)
```

```
## [1] 2
```

```
add_2(1)
```

```
## [1] 3
```

No problems here, but now consider this:

```
adders <- vector("list", 3)
for (i in 1:3) adders[[i]] <- make_adder(i)
```

The intent here is to create three adder functions that add 1, 2, and 3, respectively, to their argument. When we call the first function, though, we get an unpleasant surprise.

```
adders[[1]](1)
```

```
## [1] 4
```

The expression n + m inside the closure is not evaluated until we call it. Before we evaluate the body in the adders[[1]](1) call, the environment graph looks like Figure 7-5. All three adders are closures that refer to different instances of make_adders, but all these instances have n refer to a promise that is the expression i. The variable i is found in the global environment and not in the closure environment. After we have created all three closures, i refers to the number 3. To evaluate n + m inside the adder, we must first evaluate the promise that n refers to. We search for n and find it in the parent environment of the function call (the closure environment) where n refers to i that should be evaluated in the global environment. We evaluate it and now n refers to 3, as shown in Figure 7-6. This is why the result of calling adders[[1]] with m set to 1 returns 4 and not 2.

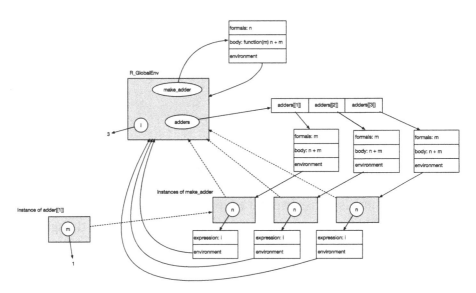

Figure 7-5. *Adders before evaluating the body of the adders[[1]](1) call*

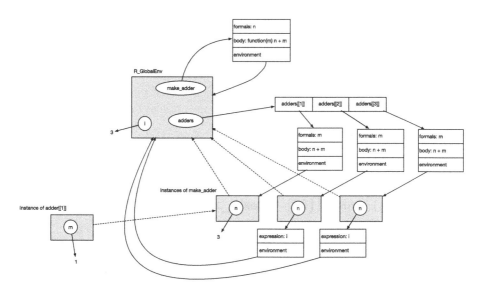

Figure 7-6. *Adders after evaluating the n promise in adders[[1]]*

After we have called this closure, the variable n no longer refers to a promise but to the value 3, so changing i at this point will not affect the closure.

```
i <- 1
adders[[1]](1)
```

```
## [1] 4
```

It will, however, affect the closures where we haven't evaluated the promise yet, so if we call one of the other closures after changing i, we will see the result of the change.

```
adders[[2]](1)
```

```
## [1] 2
```

This is a problem that can occur only when you create closures, but every time you do, the risk is there. You can avoid the problem by explicitly evaluating promises before you return the closure; this is what the function force is for.

```
make_adder <- function(n) {
  force(n)
  function(m) n + m
}
for (i in 1:3) adders[[i]] <- make_adder(i)
for (i in 1:3) print(adders[[i]](0))

## [1] 1
## [1] 2
## [1] 3
```

Quotes and Non-standard Evaluation

What we have seen so far in this chapter is the standard way to evaluate expressions, but as you can probably guess, the reason we call it the standard way is because there are alternatives to it—non-standard evaluation. That would be any other way we could evaluate expressions.

Non-standard evaluation follows the same rules from looking up variables to value mappings that standard evaluation follows. We have a chain of environments, and we search them in turn. What makes it non-standard evaluation is that we chain together environments in alternative ways.

To implement non-standard evaluation, we first need an expression to evaluate—rather than the value that is the result of evaluating one. We have already seen two ways of obtaining such an expression: we have used quote to get an expression from a literal expression, or we can use substitute to translate a function argument into an expression. There are other ways to create quoted expressions—see, for example, functions expression and bquote—and substitute can be used for more than simply translating function arguments into expressions, but quote and substitute on arguments suffice for most uses of non-standard evaluation. They both give us a quoted expression with no environment associated with it.

```
ex1 <- quote(2 * x + y)
ex1
```

```
## 2 * x + y
```

```
f <- function(ex) substitute(ex)
ex2 <- f(2 * x + y)
ex2
```

```
## 2 * x + y
```

When implementing lambda expressions, we used such expressions to create new functions.

```
g <- rlang::new_function(alist(x=, y=), body = ex1)
g
```

```
## function (x, y)
## 2 * x + y
```

```
g(1,3)
```

```
## [1] 5
```

A more direct way to evaluate an expression is using eval.

```
x <- 1
y <- 3
eval(ex1)
```

```
## [1] 5
```

With eval, we will evaluate the expression in the environment where we call eval by default, so previously we evaluated ex1 in the global environment, and in the following example we evaluate it in the local environment of calls to function h.

```
h <- function(x, y) eval(ex1)
h
```

```
## function(x, y) eval(ex1)
```

```
h(1,3)
```

```
## [1] 5
```

If we use the default environment in calls to eval, we get the standard evaluation, but we do not *have* to use the default environment. We can provide an environment to eval, which is the one we want it to evaluate the expression in. For example, we can make function h evaluate ex1 in the calling environment instead of its own local environment.

```
h <- function(x, y) eval(ex1, rlang::caller_env())
x <- y <- 1
h(4,4)
```

```
## [1] 3
```

Here, we call h from the global environment where x and y are set to 1. Even though the local variables in the call to h are 4 and 4, 2 * x + y evaluates to 3 because it is the values of x and y in the global environment that are used.

Similarly, we can use an alternative environment for functions we create. By default, new_function will use the environment where we create the function, so for example, we can create a function that creates a closure this way:

```
f <- function(x) rlang::new_function(alist(y=), ex1)
f(2)
```

```
## function (y)
## 2 * x + y
## <environment: 0x7fdd819805f8>
```

f(2)(2)

```
## [1] 6
```

We can provide an environment to new_function, however, to change this behavior. Consider, for example, this function:

```
g <- function(x) {
  rlang::new_function(alist(y=), ex1, rlang::caller_env())
}
g(2)
```

```
## function (y)
## 2 * x + y
```

g(2)(2)

```
## [1] 4
```

When we call g, we get a new function, but *this* function will be evaluated in the scope where we *call* g, not the scope *inside* the call to g. Thus, the argument x to g will not be used when evaluating 2 * x + y. In this example, we instead use the global variable x, which we set to 1 earlier.

With eval, the environment parameter doesn't have to be an environment. You can use a list or a data.frame (which is strictly speaking also a list) instead.

eval(ex1, **list**(x = 4, y = 8))

```
## [1] 16
```

df <- **data.frame**(x = 1:4, y = 1:4)
eval(ex1, df)

```
## [1]  3  6  9 12
```

Evaluating expressions in the scope of lists and data frames is a powerful tool exploited in domain-specific languages such as dplyr. But lists and data frames do not have the graph structure that environments have, which leads us to ask: if we do not find a variable in the list or data frame, where do we find it when we call eval? To answer this, eval takes a third argument that determines the enclosing scope. If variables are not found in the environment parameter, then eval will search in the enclosing scope parameter.

Consider the functions f and g defined here:

```
f <- function(expr, data, y) eval(expr, data)
g <- function(expr, data, y) eval(expr, data, rlang::caller_env())
```

They both evaluate an expression in a context defined by data, but f then uses the function call scope as the enclosing scope, while g uses the calling scope as the enclosing environment in the call to eval. Both take the parameter y, but if we use y in the expression we pass to the functions, only f will use the parameter; g, on the other hand, will look for y in the calling scope if it is not in data.

```
df <- data.frame(x = 1:4)
y <- 1:4
f(quote(x + y), df, y = 5:8) == 1:4 + 5:8

## [1] TRUE TRUE TRUE TRUE

g(quote(x + y), df, y = 5:8) == 1:4 + 1:4

## [1] TRUE TRUE TRUE TRUE
```

The combination of quoted expressions and non-standard evaluation is undoubtedly a powerful tool for creating domain-specific languages. However, it has its pitfalls: complications on who is responsible for quoting expressions and complications on stringing environments together correctly.

Let's consider these in turn. Some code must be responsible for turning an expression into a quoted expression. The simplest solution to this is to leave it up to the user to always quote expressions that must be quoted. This would be the solution in a function like this:

```
f <- function(expr, data) eval(expr, data, rlang::caller_env())
f(quote(u + v), data.frame(u = 1:4, v = 1:4))
```

```
## [1] 2 4 6 8
```

It is, however, a bit cumbersome to explicitly quote every time you call such a function, and it goes against the spirit of domain-specific languages where we want to make new syntax for easier code writing. However, if we let the function quote the expression using substitute, as in this function:

```
fq <- function(expr, data) {
  eval(substitute(expr), data, rlang::caller_env())
}
fq(u + v, data.frame(u = 1:4, v = 1:4))
```

```
## [1] 2 4 6 8
```

then we potentially run into problems if we want to call this function from another function. We can try just calling fq with an expression.

```
g <- function(expr) fq(expr, data.frame(u = 1:4, v = 1:4))
g(u + v)
```

```
## Error in eval(substitute(expr), data, rlang::caller_env()):
object 'u' not found
```

This doesn't work because expr is now considered a promise that should be evaluated in the global scope, so inside fq we try to evaluate the expression, which we cannot do because u and v are not defined. We would be even worse off if we used an expression that we actually *can* evaluate because it wouldn't be obvious that we were evaluating it in the wrong scope and thus on the wrong data.

```
u <- v <- 5:8
g(u + v)
```

```
## [1] 10 12 14 16
```

We could try to get the expression quoted using substitute inside g.

```
g <- function(expr) {
  fq(substitute(expr), data.frame(u = 1:4, v = 1:4))
}
g(u + v)
```

```
## expr
```

This fails in a different way. The expression that we get inside fq when that function calls substitute is the expression the function was called with, which is substitute(expr). So, it evaluates substitute(substitute(expr)) and gets expr, not u + v. The same would happen if we used quote, in this case because quote(expr) doesn't substitute the function argument into expr.

```
g <- function(expr) {
  fq(quote(expr), data.frame(u = 1:4, v = 1:4))
}
g(u + v)
```

```
## expr
```

There is no good way to resolve this problem. If you call a function that quotes an expression, you should give it a literal expression to quote. Such functions are essentially not useful for programming—they provide an interface to a user of your domain-specific language, but you cannot use them to implement the language by calling them from other functions.

The solution is to have functions that expect expressions to be quoted, like the function f we wrote before fq, and use them when you call one function from another.

```
g <- function(expr) {
  f(substitute(expr), data.frame(u = 1:4, v = 1:4))
}
g(u + v)
```

```
## [1] 2 4 6 8
```

If you want some functionality to be available for programming—in other words, calling a function from another function—and also as an operation in your language, then write one that expects expressions to be quoted and another that wraps it.

```
f <- function(expr, data) eval(expr, data, rlang::caller_env())
fq <- function(expr, data) f(substitute(expr), data)
fq(u + v, data.frame(u = 1:4, v = 1:4))
```

```
## [1] 2 4 6 8
```

This, however, brings us to the second pitfall—getting environments wired up correctly. Consider these two functions:

```
g <- function(x, y, z) {
  w <- x + y + z
  f(quote(w + u + v), data.frame(u = 1:4, v = 1:4))
}
h <- function(x, y, z) {
  w <- x + y + z
  fq(w + u + v, data.frame(u = 1:4, v = 1:4))
}
```

Function g explicitly quotes the expression w + u + v and calls f; h instead calls fq that takes care of the quoting for it. The first function works, the second does not.

```
g(1:4, 1:4, 1:4) == (1:4 + 1:4 + 1:4) + 1:4 + 1:4
```

```
## [1] TRUE TRUE TRUE TRUE
```

```
h(1:4, 1:4, 1:4) == (1:4 + 1:4 + 1:4) + 1:4 + 1:4
```

```
## Error in eval(expr, data, rlang::caller_env()): object 'w'
not found
```

This time, the problem is not quoting. Both functions attempt to evaluate the same expression, w + u + v, inside function f. The problem is that the variable w is available to f only when we call it from g. To see why, consider the environments in play. We do not define any nested functions, so all four functions (f, fq, g, and h) only have access to their local environment and the global environment. The expression that f gets as its argument, however, is not evaluated in f's local environment but in its caller's environment. When f is called directly from g, the caller environment is the local environment of the g call, where w is defined. When f is called from h, however, it is not called directly. Since h calls fq that then calls f, the caller of f in this case is fq. The variable w is defined in the local scope of h, but this is not where f tries to evaluate the expression; f tries to evaluate the expression in the scope of fq where w is *not* defined.

It is less obvious how we should resolve this issue. It is possible to pass environments along with expressions as separate function parameters, but this becomes cumbersome if we have to work with more than one expression. What we want is to associate expressions with the environment in which we want to look up variables we do not explicitly override, for example by getting them from a data frame.

Expressions do not carry along with them any environment, so we cannot get there directly. Formulas, however, do. Instead of using expressions, we can use one-sided formulas. Quoting would now involve making a formula out of an expression. If the formula is one-sided, we can get the expression as the second element in it, and the environment where the formula is defined is available using the environment function. We can rewrite the f and fq functions to be based on formulas.

```
ff <- function(expr, data) {
  eval(expr[[2]], data, environment(expr))
}
ffq <- function(expr, data) {
  expr <- eval(substitute(~ expr))
  environment(expr) <- rlang::caller_env()
  ff(expr, data)
}
```

With ff you need to explicitly create the formula—similar to how you had to quote expressions in f explicitly—and this automatically gives you the environment associated with the formula. With ffq we translate an expression into a formula using substitute and explicitly set its environment to the caller environment. We can now define g and h similar to before, except that g uses a formula instead of quote.

```
g <- function(x, y, z) {
  w <- x + y + z
  ff(~ w + u + v, data.frame(u = 1:4, v = 1:4))
}
h <- function(x, y, z) {
  w <- x + y + z
  ffq(w + u + v, data.frame(u = 1:4, v = 1:4))
}
```

This time, both functions will evaluate the expressions in the right scope.

```
g(1:4, 1:4, 1:4) == (1:4 + 1:4 + 1:4) + 1:4 + 1:4
```

```
## [1] TRUE TRUE TRUE TRUE
```

```
h(1:4, 1:4, 1:4) == (1:4 + 1:4 + 1:4) + 1:4 + 1:4
```

```
## [1] TRUE TRUE TRUE TRUE
```

Associating environments to expressions is the idea behind *quosures* from the rlang package. The word is a portmanteau created from quotes and closures—similar to how closures are functions with associated environments, quosures are quoted expressions with associated environments. Quosures are based on formulas, and we could use formulas as in the example we just saw, but the rlang package provides functionality that makes it much simpler to program domain-specific languages using quosures. The rlang package implements so-called tidy evaluation, which is the topic of the next chapter.

CHAPTER 8

Tidy Evaluation

The so-called tidyverse refers to a number of R packages designed to work well together and based on similar designs that can all be considered domain-specific languages in themselves. These packages include `dplyr`, `tidyr`, and `ggplot2` and mainly consist of functions that do non-standard evaluation. The way they manage non-standard evaluation is consistent among the packages and based on what they call *tidy evaluation*, which primarily relies on two features implemented in the `rlang` package: quosures and quasi-quotation.

We will use the `rlang` package once more, but to make it explicit when we use this package and not basic R non-standard evaluation, I will usually use fully qualified names. In other words, I will write `rlang::quo` instead of `quo` and not load the package. We will also use the `purrr` package, but here as well I will use fully qualified names. I *will* load `magrittr` to get the pipeline operator, however, and the `tibble` and `dplyr` packages for working with data frames.

```
library(magrittr)
library(dplyr)
library(tibble)
```

© Thomas Mailund 2018
T. Mailund, *Domain-Specific Languages in R*,
https://doi.org/10.1007/978-1-4842-3588-1_8

Quosures

The rlang package provides functions to replace quote and substitute that create quosures—expressions based on formulas that carry with them their environment—instead of quoted expressions. To create a quosure from an expression, you use quo.

```
q <- rlang::quo(2 * x)
q
```

```
## <quosure>
##   expr: ^2 * x
##   env:  global
```

Inside a function call, the quosure analog to substitute is enquo.

```
f <- function(expr) rlang::enquo(expr)
q <- f(2 * x)
q
```

```
## <quosure>
##   expr: ^2 * x
##   env:  global
```

In both examples, the scope associated with the quosure is the global environment because this is the level at which the expression is written.

A quosure is just a special type of formula, so we can access one as we did in the previous section to get the environment and expression,

```
q[[2]]
```

```
## 2 * x
```

```
environment(q)
```

```
## <environment: R_GlobalEnv>
```

However, `rlang` provides functions for working with quosures that make the intent of our code clearer. To get the expression out of a quosure, we use the function get_expr.

```
rlang::get_expr(q)
```

```
## 2 * x
```

The term is ironic since the *un*quote function returns a quoted expression, but it does strip away the quosure-ness and gives us a raw expression. There is another unquote function, UQ, that *does* unquote an expression in the sense of evaluating it, but it has a different purpose that we get to in the next section.

Getting the environment associated with a quosure can be done using environment as we saw earlier, but the `rlang` function for this is get_env.

```
rlang::get_env(q)
```

```
## <environment: R_GlobalEnv>
```

With quosures, you can no longer evaluate them with eval. A quosure is a formula, and the result of evaluating a formula is the formula itself.

```
eval(q)
```

```
## <quosure>
##   expr: ^2 * x
##   env:  global
```

Instead, you need to use the function eval_tidy.

```
x <- 1
rlang::eval_tidy(q)
```

```
## [1] 2
```

```
x <- 2
```

```
rlang::eval_tidy(q)
```

```
## [1] 4
```

The quosure is evaluated inside the environment it is associated with. We created quosure q inside function f, but its environment is the global environment, so when we modify this, by changing x, it affects the result of evaluating q.

If we create a quosure inside a local function scope, it will remember this context—just like a closure. For example, if we define function f as this

```
f <- function(x, y) rlang::quo(x + y + z)
```

the quosure will know the function parameters x and y from when f is called but will have to find z elsewhere. Consider the contrast between evaluating the quosure and the expression x + y + z directly, shown here:

```
q <- f(1, 2)
x <- y <- z <- 3
rlang::eval_tidy(q) # 1 + 2 + 3
```

```
## [1] 6
```

```
x + y + z # 3 + 3 + 3
```

```
## [1] 9
```

Just like eval, eval_tidy lets you provide a list, data frame, or environment with bindings from variables to values. When you do this, the values you provide will overrule the variables in the quosure's environment—an effect known as *over-scoping*. Consider this:

```
x <- 1:4
y <- 1:4
q <- quo(x+y)
```

```
rlang::eval_tidy(q)
```

```
## [1] 2 4 6 8
```

```
rlang::eval_tidy(q, list(x = 5:8))
```

```
## [1]  6  8 10 12
```

The quosure q is bound to the global environment, so when we evaluate it, x and y are both 1:4. However, when we provide the second argument to eval_tidy, we can override the value of x to 5:8. You will recognize this feature from dplyr where you have access to columns in data frames in arguments you provide to the functions there, and these columns overrule any global variable that might otherwise have been used.

This can also be used to override variables in a function call with function parameters. Consider these two functions:

```
f <- function(expr,x) {
  q <- rlang::enquo(expr)
  rlang::eval_tidy(q)
}
g <- function(expr,x) {
  q <- rlang::enquo(expr)
  rlang::eval_tidy(q, environment())
}
f(x + y, x = 5:8)
```

```
## [1] 2 4 6 8
```

```
g(x + y, x = 5:8)
```

```
## [1]  6  8 10 12
```

The function f evaluates the quosure in its scope, which doesn't contain the function parameter x, while the function g over-scopes with the function environment. This makes the variable x refer to the function parameter rather than the global parameter.

The expression you evaluate with eval_tidy doesn't have to be a quosure. The function is equally happy to evaluate bare expressions, and then it behaves just like eval.

```
rlang::eval_tidy(quote(x + y))
```

```
## [1] 2 4 6 8
```

Just like eval, eval_tidy takes a third argument that will behave as the enclosing scope. This is used for bare expressions—those created with quote.

```
rlang::eval_tidy(quote(xx), env = list2env(list(xx = 5:8)))
```

```
## [1] 5 6 7 8
```

It is not used with quosures.

```
rlang::eval_tidy(quo(xx), env = list2env(list(xx = 5:8)))
```

```
## Error in rlang::eval_tidy(quo(xx), env = list2env(list
(xx = 5:8))): object 'xx' not found
```

The list2env function I have used here translates a list into an environment—as strongly hinted by the name. It is a quick way to construct an environment and populate it with variables.

If you want to create a closure with over-scoping (in other words, you want to create a function that evaluates a quosure where it first finds local variables and then look in the quosure's environment), you cannot directly call eval_tidy when creating the function. This would ask R to attempt to evaluate the closure, but you do not yet have the variables you need—

those are provided when you call the closure. Instead, you can separate the bare expression and the environment of the quosure using get_expr and get_env, respectively. Consider the function make_function shown here:

```
make_function <- function(args, body) {
  body <- rlang::enquo(body)
  rlang::new_function(args, rlang::get_expr(body), rlang::get_
  env(body))
}
f <- function(z) make_function(alist(x=, y=), x + y + z)
g <- f(z = 1:4)
g

## function (x, y)
## x + y + z
## <environment: 0x7f9b0154b808>

g(x = 1:4, y = 1:4)

## [1]  3  6  9 12
```

Here, make_function takes two arguments, a pair list of arguments and an expression for the body of a function. It is slightly more primitive than the lambda expressions we wrote in the previous chapter, but it is essentially doing the same thing. In this example, I am focusing on the closure we create rather than on language design issues. We translate the function body into a quosure, which guarantees that we have an environment associated with it. In the function we create using new_ function, however, we strip the environment from the quosure to create the body of the new function, but we assign the function's environment to be the quosure environment.

In the function f we have a local scope that knows the value of z, and we create a new function in this scope. The quosure we get from this is associated with the local f scope, so it also knows about z. We provide variables x and y, when calling g, but the z value is taken from the local scope of f.

If called directly, there is no difference between using the caller's environment or the quosure's environment.

```
make_function_quo <- function(args, body) {
  body <- rlang::enquo(body)
  rlang::new_function(args, rlang::get_expr(body), rlang::get_
  env(body))
}
make_function_quote <- function(args, body) {
  body <- substitute(body)
  rlang::new_function(args, body, rlang::caller_env())
}
g <- make_function_quo(alist(x=, y=), x + y)
h <- make_function_quote(alist(x=, y=), x + y)
g(x = 1:4, y = 1:4)

## [1] 2 4 6 8

h(x = 1:4, y = 1:4)

## [1] 2 4 6 8
```

However, consider a more involved example, where we collect expressions in a list and have a function for translating all the expressions into functions that we can then apply over values using the invoke_map function from the purrr package. We can construct the expressions like this, using the linked lists structure we have previously used:

```
cons <- function(elm, lst) list(car=elm, cdr=lst)
lst_length <- function(lst) {
  len <- 0
```

```
  while (!is.null(lst)) {
    lst <- lst$cdr
    len <- len + 1
  }
  len
}
lst_to_list <- function(lst) {
  v <- vector(mode = "list", length = lst_length(lst))
  index <- 1
  while (!is.null(lst)) {
    v[[index]] <- lst$car
    lst <- lst$cdr
    index <- index + 1
  }
  v
}
expressions <- function() list(ex = NULL)
add_expression <- function(ex, expr) {
  ex$ex <- cons(rlang::enquo(expr), ex$ex)
  ex
}
```

Translating the expressions into functions is straightforward. We need to reverse the resulting list only if we want the functions in the order we add them since we prepend expressions when we use the linked lists.

```
make_functions <- function(ex, args) {
  results <- vector("list", length = lst_length(ex$ex))
  i <- 1; lst <- ex$ex
  while (!is.null(lst)) {
    results[[i]] <-
```

```
    rlang::new_function(args, rlang::get_expr(lst$car),
                        rlang::get_env(lst$car))
    i <- i + 1
    lst <- lst$cdr
  }
  rev(results)
}
```

With this small domain-specific language for collecting expressions, we can write a function that creates expressions for computing y-coordinates of a line given an intercept.

```
make_line_expressions <- function(intercept) {
  expressions() %>%
    add_expression(coef + intercept) %>%
    add_expression(2*coef + intercept) %>%
    add_expression(3*coef + intercept) %>%
    add_expression(4*coef + intercept)
}
```

The expressions know the intercept when we call make_line_expressions—that is the intent at least—but the coefficient should be added later in a function call. We can create the functions for the expressions using another function.

```
eval_line <- function(ex, coef) {
  ex %>% make_functions(alist(coef=)) %>%
    purrr::invoke_map(coef = coef) %>% unlist()
}
```

The invoke_map function is similar to the various map functions from purrr, but instead of mapping a function over several values, it takes a sequence of functions and applies each to a value.

We can now pipe these functions together to get points on a line.

```
make_line_expressions(intercept = 0) %>% eval_line(coef = 1)
## [1] 1 2 3 4
make_line_expressions(intercept = 0) %>% eval_line(coef = 2)
## [1] 2 4 6 8
make_line_expressions(intercept = 1) %>% eval_line(coef = 1)
## [1] 2 3 4 5
```

Everything works as intended here, but what would happen if we used quotes instead? It is simple to write the corresponding functions.

```
add_expression <- function(ex, expr) {
  ex$ex <- cons(substitute(expr), ex$ex)
  ex
}
make_functions <- function(ex, args) {
  results <- vector("list", length = lst_length(ex$ex))
  i <- 1; lst <- ex$ex
  while (!is.null(lst)) {
    results[[i]] <- rlang::new_function(args, lst$car,
    rlang::caller_env())
    i <- i + 1
    lst <- lst$cdr
  }
  rev(results)
}
```

We will get an error if we try to use them as before, however.

```
make_line_expressions(intercept = 0) %>% eval_line(coef = 1)
## Error in (function (coef) : object 'intercept' not found
```

The reason for this is obvious once we consider which environments contain information about the intercept. This variable lives in the scope of calls to make_line_expressions, but when we create the functions, we do so by calling make_functions from inside eval_line. The functions are created with eval_line local environments as their closures, and intercept is not found there.

In general, it is safer to use quosures than bare expressions for non-standard evaluation exactly because they carry their environment with them, alleviating the problems we have with keeping track of which environment to evaluate expressions in.

One thing to mention before we move on to the next topic is the function quos. This function works as quo but for a sequence of arguments that are returned as a list of quosures.

```
rlang::quos(x, y, x+y)
```

```
## [[1]]
## <quosure>
##   expr: ^x
##   env:  global
##
## [[2]]
## <quosure>
##   expr: ^y
##   env:  global
##
## [[3]]
## <quosure>
##   expr: ^x + y
##   env:  global
```

The primary use of quos is to translate the three-dots argument into a list of quosures.

```
f <- function(...) rlang::quos(...)
f(x, y, z)

## [[1]]
## <quosure>
##   expr: ^x
##   env:  global
##
## [[2]]
## <quosure>
##   expr: ^y
##   env:  global
##
## [[3]]
## <quosure>
##   expr: ^z
##   env:  global
```

Quasi-quoting

The final topic of this chapter involves *quasi-quoting*. This is a mechanism by which we can work with quoted expressions but at the same time substitute some parts of the expression by the value that subexpressions evaluate to. When we directly call functions that do non-standard evaluation, we can usually provide expressions exactly as we want them, but as soon as we start using such non-standard evaluation functions in programs where we call them from other functions, we need some flexibility in how we construct expressions. We can do this with meta-programming where we modify call objects, but a better approach is implemented in the rlang package, the so-called quasi-quoting.

Consider this simple example. We have a data frame, and we want to filter away rows where a given column has missing data. We can do this using dplyr's filter function like this:

```
df <- tribble(
  ~x, ~y,
   1,  1,
  NA,  2,
   3,  3,
   4, NA,
   5,  5,
  NA,  6,
   7, NA
)
df %>% dplyr::filter(!is.na(x))

## # A tibble: 5 x 2
##       x     y
##    <dbl> <dbl>
## 1    1.    1.
## 2    3.    3.
## 3    4.    NA
## 4    5.    5.
## 5    7.    NA

df %>% dplyr::filter(!is.na(y))

## # A tibble: 5 x 2
##       x     y
##    <dbl> <dbl>
## 1    1.    1.
## 2    NA    2.
## 3    3.    3.
## 4    5.    5.
## 5    NA    6.
```

Here, we use the same code for two different variables. It is straightforward code, so we would not write a function to avoid the duplication, but for more complicated pipelines, we would. Therefore, for the sake of argument, let's imagine that we want to replace the pipeline with a function. We could attempt to write it like this:

```
filter_on_na <- function(df, column) {
  column <- substitute(column)
  df %>% dplyr::filter(!is.na(column))
}
df %>% filter_on_na(x)

## Warning in is.na(column): is.na() applied to non-
## (list or vector) of type 'symbol'

## # A tibble: 7 x 2
##       x      y
##    <dbl>  <dbl>
## 1    1.     1.
## 2    NA     2.
## 3    3.     3.
## 4    4.     NA
## 5    5.     5.
## 6    NA     6.
## 7    7.     NA
```

We use substitute to translate the column name into a symbol and then apply the filter pipeline. It doesn't work, of course, and the reason is that filter does non-standard evaluation as well and translates the predicate !is.na(column) into this exact expression. So, it needs to know the variable column. Now, since filter evaluates its argument as a quosure, it *can* find the variable column, but it finds that this is a symbol—it is, in this case, the symbol x—but that is not what we want it to see. We want filter to see the column x, but that is not how filter works.

What we want to do is to substitute the symbol held in the variable column into the expression/quosure that filter sees. We can do this with the "bang-bang" operator !!.

```
filter_on_na <- function(df, column) {
  column <- rlang::enexpr(column)
  df %>% dplyr::filter(!is.na(!!column))
}
df %>% filter_on_na(x)

## # A tibble: 5 x 2
##       x     y
##    <dbl> <dbl>
## 1    1.    1.
## 2    3.    3.
## 3    4.    NA
## 4    5.    5.
## 5    7.    NA

df %>% filter_on_na(y)

## # A tibble: 5 x 2
##       x     y
##    <dbl> <dbl>
## 1    1.    1.
## 2    NA    2.
## 3    3.    3.
## 4    5.    5.
## 5    NA    6.
```

This operator unquotes the following expression, evaluates it, and puts the result into the quoted expression that filter sees. So the value of column, rather than the symbol column, gets inserted into !is. na(!!column).

You will have noticed that I used a different function, `rlang::enexpr`, to create the quoted column name. In the `rlang` package there are two functions that behave like `quote` and `substitute` in that they create bare expressions, but they allow quasi-quoting with the bang-bang operator, something `substitute` and `quote` do not. Consider the functions f and g defined like this:

```
f <- function(x) substitute(x)
g <- function(x) rlang::enexpr(x)
```

When called directly, both will return the expression we give as the parameter x, but consider the case where we call them from another function, h, where we want to substitute its parameter for the parameter x.

```
h <- function(func, var) func(!!var)
h(f, quote(x))

## !(!var)

h(g, quote(x))

## x
```

In f, where we use `substitute`, we get the expression `!!var` back, but in function g, where we use `enexpr`, we get the substitution done and get the desired result x.

The function enexpr works like enquo to translate function parameters into expressions, but it translates them into bare expressions rather than quosures. As the analog to quo, which takes an expression directly and creates a quosure, we have expr, which creates a bare expression—like quote—but allows quasi-quotation.

```
x <- y <- 1
quote(2 * x + !!y)

## 2 * x + (!(!y))

rlang::expr(2 * x + !!y)
```

```
## 2 * x + 1

rlang::quo(2 * x + !!y)

## <quosure>
##   expr: ^2 * x + 1
##   env:  global
```

You have to be a little bit careful when using the bang-bang operator, depending on which version of the rlang package you use. It is built from the negation operator, !, which has a low precedence. The only operators with lower precedence are the logical operators and assignments. This means if you try to unquote x in the expression x + y, you might, in fact, be unquoting the entire expression if you simply write !!x + y. You only get !! bound to x if the operator you use in the expression is a logical operator.

If you try to evaluate this express, you will see whether you have an old version or a new version of the package. If you have an old version, the first expression will evaluate x + y since ! has lower precedence than +; if you have a newer version of rlang, then only the value of x will be substitute into the expression.

```
x <- y <- 2
rlang::expr(!!x + y)

## 2 + y
```

The most recent version of the package, at the time of writing, has resolved this; instead of relying on R's precedence rules, it will examine the expression before it evaluates unquoted expressions, and it gives the bang-bang operator a much tighter precedence than the negation operator has. If you do not have the latest version of the package, however, you can get around this problem using parentheses, or you can use the function-variant of unquoting, UQ.

```
rlang::expr(UQ(x) + y)
```

```
## 2 + y
```

When we translate a function argument into an expression, with enexpr, or a quosure, with enquo, the bang-bang operator or the UQ function will substitute the results of expressions into the expressions/quosures we create, which is why the second filter_on_na function worked.

You can unquote on the left-hand side of named parameters in function calls as well, but the R parser does not allow you to write code such as this:

```
f(UQ(var) = expr)
```

This is because the left-hand side in function call arguments must be symbols. To get around this problem, the rlang package provides an implementation of the := operator that does allow such assignments. Consider the following:

```
f <- function(df, summary_name, summary_expr) {
  summary_name <- rlang::enexpr(summary_name)
  summary_expr <- rlang::enquo(summary_expr)
  df %>% mutate(UQ(summary_name) := UQ(summary_expr))
}
tibble(x = 1:4, y = 1:4) %>% f(z, x + y)
```

```
## # A tibble: 4 x 3
##       x     y     z
##    <int> <int> <int>
## 1     1     1     2
## 2     2     2     4
## 3     3     3     6
## 4     4     4     8
```

The named function call in the following:

```
df %>% mutate(UQ(summary_name) := UQ(summary_expr))
```

behaves like a usual function call with named parameters after the quasi-quotation substitution has taken place.

A final quasi-quotation function you must know is UQS. This function behaves as UQ in that it evaluates its arguments and puts the result into a quoted expression, but it is intended for splicing a list of quosures into a function call.

Consider this example:

```
args <- rlang::quos(x = 1, y = 2)
q <- rlang::expr(f(rlang::UQS(args)))
q

## f(x = ~1, y = ~2)
```

Here, we create a list of quosures for arguments x and y and create an expression that calls the function f with these arguments. The UQS function is what splices the arguments into the expression for the function call. Using UQ, we would get a different expression.

```
q <- rlang::expr(f(rlang::UQ(args)))
q

## f(list(x = ~1, y = ~2))
```

As for UQ, though, there is an operator version of UQS, the triple-bang operator: !!!.

```
q <- rlang::expr(f(!!!args))
q

## f(x = ~1, y = ~2)
```

To see UQS in action, we consider another toy example. We write a function for evaluating the mean of an expression given a data frame. Imagine that we want to map over several data frames and compute the mean of the same expression or something like that. We construct a function for this that evaluates an expression we capture as a quosure and computes the mean of the expression in a data frame, using additional arguments to modify the call to mean.

```r
mean_expr <- function(ex, ...) {
  ex <- rlang::enquo(ex)
  extra_args <- rlang::dots_list(...)
  mean_call <- rlang::expr(with(
      data,
      mean(!!rlang::get_expr(ex), !!!extra_args))
  )
  rlang::new_function(args = alist(data=),
                      body = mean_call,
                      env = rlang::get_env(ex))
}
mean_sum <- mean_expr(x + y, na.rm = TRUE)
mean_sum

## function (data)
## with(data, mean(x + y, na.rm = TRUE))
```

There are a few new things to observe here. We use dots_list to evaluate the arguments in the ... parameter. We could deal with this in other ways, but we don't necessarily want them to be evaluated lazily, and we just want a list to use as extra arguments to a call to mean. We create an expression from the ex parameter. Here, we wrap the bare expression in ex inside an expression that involves the with function to include data in the form of a data frame. We use the !!! operator to splice the additional parameters into the call to mean. We then construct a new function that takes a single argument, data; evaluates the with(mean(...)) expression in its body; and is evaluated in the scope where ex was defined.

We can test it with the data frame we created earlier:

```
df
```

```
## # A tibble: 7 x 2
##        x     y
##    <dbl> <dbl>
## 1    1.    1.
## 2    NA    2.
## 3    3.    3.
## 4    4.    NA
## 5    5.    5.
## 6    NA    6.
## 7    7.    NA
```

```
mean_sum(df)
```

```
## [1] 6
```

To see that the environment we evaluated the expression in is where it was defined, we can write a function to capture a parameter in a local scope and see that this scope is available when we call the mean function.

```
f <- function(z) mean_expr(x + y + z, na.rm = TRUE, trim = 0.1)
g <- f(z = 1:7)
g
```

```
## function (data)
## with(data, mean(x + y + z, na.rm = TRUE, trim = 0.1))
## <environment: 0x7f9b02d4ab88>
```

```
g(df)
```

```
## [1] 9
```

Using quosures and quasi-quotes adds to the problems you can have with keeping track of environments for non-standard evaluation. I strongly suggest you use these instead of the bare R quote and substitute approach. They do not, however, fix the problem with deciding when to quote arguments and with calling functions that translate arguments into expressions. If you call a function that uses enquo from a function where you have already quoted the argument, then you get it double-quoted.

```
f <- function(expr) rlang::enquo(expr)
g <- function(expr) f(rlang::enquo(expr))
f(x + y)

## <quosure>
##    expr: ^x + y
##    env:  global

g(x + y)

## <quosure>
##    expr: ^rlang::enquo(expr)
##    env:  0x7f9b016be9c8

rlang::eval_tidy(f(x + y), list(x = 1, y = 2))

## [1] 3

rlang::eval_tidy(g(x + y), list(x = 1, y = 2))

## <quosure>
##    expr: ^x + y
##    env:  global
```

Here, the solution is as before: if you need to call functions with already quoted expressions, make two versions—one that expects its argument to be quoted and one that does it for you.

CHAPTER 9

List Comprehension

We will now use what we have learned to implement a valuable language construction that is not built into R: *list comprehension*. List comprehensions provide a syntax for mapping and filtering sequences. In R we would use functions such as Map or Filter, or the purrr alternatives, for this, but in languages such as Haskell or Python, there is syntactic sugar to make combinations of mapping and filtering easier to program.

Take an algorithm such as quicksort. Here, the idea is to sort a list by picking a random element in it, called the *pivot*, splitting the data into those elements smaller than the pivot, equal to the pivot, and larger than the pivot. We then sort those smaller and larger elements recursively and concatenate the three lists to get the final sorted list. One way to implement this in R is to use the Filter function.

```r
qsort <- function(lst) {
  n <- length(lst)
  if (n < 2) return(lst)

  pivot <- lst[[sample(n, size = 1)]]
  smaller <- Filter(function(x) x < pivot, lst)
  equal <- Filter(function(x) x == pivot, lst)
  larger <- Filter(function(x) x > pivot, lst)
  c(qsort(smaller), equal, qsort(larger))
}
```

© Thomas Mailund 2018
T. Mailund, *Domain-Specific Languages in R*,
https://doi.org/10.1007/978-1-4842-3588-1_9

```
(lst <- sample(1:10))
```

```
##  [1]  3  7  8  2  1  4 10  5  9  6
```

```
unlist(qsort(lst))
```

```
##  [1]  1  2  3  4  5  6  7  8  9 10
```

This is readable if you are familiar with functional programming, but it does take some decoding to work out the Filter expression and decode the predicate used in it. Compare this to a Python implementation that does the same thing (except that the pivot is not chosen randomly because sampling is required in Python).

```python
def qsort(lst):
    if len(lst) < 2:
        return lst
    pivot = lst[0]
    return qsort([x for x in lst if x < pivot]) +
                 [x for x in lst if x == pivot] +
           qsort([x for x in lst if x > pivot])
```

Or consider a similar Haskell implementation, shown here:

```haskell
qsort lst =
    if length lst < 2 then
        lst
    else
        let pivot = lst !! 0
        in qsort([x | x <- lst, x < pivot]) ++
                 [x | x <- lst, x == pivot] ++
           qsort([x | x <- lst, x > pivot])
```

Expressions such as the following in Python:

```python
[x for x in lst if x < pivot]
```

or the following in Haskell:

```
[x | x <- lst, x < pivot]
```

is what we call *list comprehension*. List comprehensions consist of three components, first an expression that will be evaluated for each element in the list (or lists if we use more than one), then one or more lists to map over, and finally zero or more predicates we use to filter over. It is thus a combination of Map and Filter calls in one expression.

Using non-standard evaluation, we can write an R function that provides a similar list comprehension syntax. We will write it such that its first argument must be an expression that we evaluate for all elements in the input list (or lists) and such that its remaining elements identify either lists or predicates. We will use named arguments to identify when an argument defines a list and unnamed arguments for predicates.

The function will work as follows: we take the first argument and make it into a quosure, so we have the expression plus the environment we define it in. We do the same with the rest of the arguments, captured by the three-dots parameter since we want the function to take an arbitrary number of arguments. We create the first quosure with enquo and the list of additional arguments with quos. We then split these into list arguments and predicates based on whether they are named arguments. While doing this, we evaluate the named arguments to get the data in the input lists and extract the expressions for the predicates using get_expr.

With the functions we create, both predicates and the function we use to map over the lists, we have to be a careful about which context the expression should be evaluated in. We want the expressions to be the body of functions we can map over the lists, so we can't evaluate them in the quosures' environments directly, but we do want those environments to be in scope so the expression can see variables that are not part of the list comprehension. We, therefore, get the raw expression from the quosure using the get_expr function, but functions we create from them will have the quosure environment as their enclosing scope.

We create one function per predicate and one for the main expression of the list comprehension. It is not straightforward to combine all the predicates in a filter expression to map over all the lists, but it is straightforward to use them to update a boolean vector where we keep track of which values to include in the final result. We can mask these together while applying the predicates one at a time. We can then map over the input lists and subset each of them—in the following code I use a lambda expression because these are defined in the purrr package as formulas where .x refers to the first argument. After filtering the lists, we can apply the main function over them and get the final results.

Putting all this together gives us this function:

```
library(rlang)
library(purrr)

lc <- function(expr, ...) {
  expr <- enquo(expr)
  rest <- quos(...)

  lists <- map(rest[names(rest) != ""], eval_tidy)
  predicates <- map(rest[names(rest) == ""], get_expr)

  keep_index <- rep(TRUE, length(lists[[1]]))
  for (pred in predicates) {
    p <- new_function(lists, body = pred, env = get_env(expr))
    keep_index <- keep_index & unlist(pmap(lists, p))
  }
  filtered_lists <- map(lists, ~.x[keep_index])

  f <- new_function(lists, body = get_expr(expr), env = get_
env(expr))
  pmap(filtered_lists, f)
}
```

We can use it to implement quicksort like this:

```r
qsort <- function(lst) {
  n <- length(lst)
  if (n < 2) return(lst)

  pivot <- lst[[sample(n, size = 1)]]
  smaller <- lc(x, x = lst, x < pivot)
  equal <- lc(x, x = lst, x == pivot)
  larger <- lc(x, x = lst, x > pivot)

  c(qsort(smaller), equal, qsort(larger))
}

(lst <- sample(1:10))

## [1]  9  5  7  8 10  2  1  4  3  6

unlist(qsort(lst))

## [1]  1  2  3  4  5  6  7  8  9 10
```

In this function, we only use the filtering aspects of the list comprehension, but we can use the lc function in more complex expressions. As a cute little example, we can use lc to compute the primes less than a given number n.

```r
not_primes <- lc(seq(from = 2*x, to = 100, by = x), x = 2:10) %>%
    unlist %>% unique
not_primes

## [1]   4   6   8  10  12  14  16  18  20  22  24
## [12]  26  28  30  32  34  36  38  40  42  44  46
## [23]  48  50  52  54  56  58  60  62  64  66  68
## [34]  70  72  74  76  78  80  82  84  86  88  90
```

```
## [45]  92  94  96  98 100   9  15  21  27  33  39
## [56]  45  51  57  63  69  75  81  87  93  99  25
## [67]  35  55  65  85  95  49  77  91
```

```
primes <- lc(p, p = 2:100, !(p %in% not_primes)) %>% unlist
primes
```

```
##  [1]  2  3  5  7 11 13 17 19 23 29 31 37 41 43 47
## [16] 53 59 61 67 71 73 79 83 89 97
```

This is a variant of the sieve of Eratosthenes algorithm. We compute all the numbers that are not primes (because they are multiples of the numbers), and then we identify the numbers that are not in that list. We let x go from two to 10—to identify the primes less than n it suffices to do this up to $\sqrt{n}$, and for each of those we create a list of the various multiples of x. We then get rid of duplicates to make the next step faster; in that step, we simply filter on the numbers that are not primes.

A solution for general n would look like this:

```
get_primes <- function(n) {
  not_primes <- lc(seq(from = 2*x, to = n, by = x),
   x = 2:sqrt(n)) %>%
      unlist %>% unique
  lc(p, p = 2:n, !(p %in% not_primes)) %>% unlist
}
get_primes(100)
```

```
##  [1]  2  3  5  7 11 13 17 19 23 29 31 37 41 43 47
## [16] 53 59 61 67 71 73 79 83 89 97
```

Traditionally, the algorithm doesn't create a list of non-primes first but rather starts with a list of candidates for being primes—all numbers from 2 to n. Iteratively, we then take the first element in the list, which is a prime, and remove as candidates all elements divisible by that number. We can also implement this version using a list comprehension to remove candidates:

```
get_primes <- function(n) {
  candidates <- 2:n
  primes <- NULL
  while (length(candidates) > 0) {
    p <- candidates[[1]]
    primes <- cons(p, primes)
    candidates <- lc(x, x = candidates, x %% p != 0)
  }
  primes %>% lst_to_list %>% unlist %>% rev
}
get_primes(100)

## Error in cons(p, primes): could not find function "cons"
```

As another example, where we have more than one list as input and where we use a list comprehension to construct new values rather than filter the lists, we can implement a function for zipping two lists like this:

```
zip <- function(x, y) {
  lc(c(x,y), x = x, y = y) %>% { do.call(rbind,.) }
}
zip(1:4,1:4)

##      [,1] [,2]
## [1,]    1    1
## [2,]    2    2
## [3,]    3    3
## [4,]    4    4
```

Here, we pair up elements from lists x and y in the list comprehension, and we then merge the lists using bind. The combination of do.call and bind is necessary to get a table out of this, and the curly braces are necessary to make the result of lc into the second and not the first argument of do.call. See the magrittr documentation for how curly braces are used together with the pipeline operator.

List comprehension is another example of how very little code can create a new language construct. It might be stretching it a bit to call this a language, but we *are* creating a new syntax to help us write more readable code, that is, if you consider list comprehension more readable than combinations of map and filter, of course.

## CHAPTER 10

Continuous-Time Markov Chains

We now turn to an example of a domain-specific language where we combine tidy evaluation and the `magrittr` pipe operator. We will write a language for specifying continuous-time Markov chains (CTMCs) and for computing the likelihood of parameters in such CTMCs given a trace of which states the chain is in at different time points. As with the list comprehension example in the previous chapter, we are not going to use operators to create a new syntax for the language but will create a DSL by providing functions that can be strung together to construct "sentences."

We will use the packages `magrittr` and `rlang` to construct the language, the package `tibble` for data frames, and the package `expm` for matrix exponentiation.

```r
library(magrittr)
library(rlang)
library(tibble)
library(expm)
```

We will reuse the linked list code plus the functions `collect_symbols_rec` and `make_args_list` we implemented in previous chapters.

© Thomas Mailund 2018
T. Mailund, *Domain-Specific Languages in R*,
https://doi.org/10.1007/978-1-4842-3588-1_10

```r
cons <- function(car, cdr) list(car = car, cdr = cdr)
lst_length <- function(lst) {
  len <- 0
  while (!is.null(lst)) {
    lst <- lst$cdr
    len <- len + 1
  }
  len
}
lst_to_list <- function(lst) {
  v <- vector(mode = "list", length = lst_length(lst))
  index <- 1
  while (!is.null(lst)) {
    v[[index]] <- lst$car
    lst <- lst$cdr
    index <- index + 1
  }
  v
}

collect_symbols_rec <- function(expr, lst, bound) {
  if (is.symbol(expr) && expr != "") {
    if (as.character(expr) %in% bound) lst
    else cons(as.character(expr), lst)

  } else if (is.pairlist(expr)) {
    for (i in seq_along(expr)) {
      lst <- collect_symbols_rec(expr[[i]], lst, bound)
    }
    lst

  } else if (is.call(expr)) {
    if (expr[[1]] == as.symbol("function"))
      bound <- c(names(expr[[2]]), bound)
```

```
  for (i in 1:length(expr)) {
    lst <- collect_symbols_rec(expr[[i]], lst, bound)
  }
  lst

} else {
  lst
}
}

make_args_list <- function(args) {
  res <- replicate(length(args), substitute())
  names(res) <- args
  as.pairlist(res)
}
```

We will use these functions to construct functions from a CTMC specification by extracting the unbound symbols in expressions we associate with transition rates. We will not use the collect_symbols function we implemented to collect unbound variables but instead a version that expects its expression is quoted already.

```
collect_symbols_q <- function(expr, env) {
  bound <- c()
  lst <- collect_symbols_rec(expr, NULL, bound)
  lst %>% lst_to_list() %>% unique() %>%
    purrr::discard(exists, env) %>%
    unlist()
}
```

This is because we plan to quote expressions in the DSL functions and then call this function with these quoted expressions.

Constructing the Markov Chain

We explored several approaches to design a language for CTMCs in Chapter 3. In this chapter, we will use the variation that uses the pipe operator, %>%, together with an add_edge function. We will collect edges in three lists: one list for the "from" states, one for the "to" states, and one for the rates associated with the transitions. Also, we will collect the unbound variables in the rate expressions when we create new edges, so later changes to scopes will not affect the parameters of the CTMC model. To represent a CTMC, we create a class and a list that holds the "from" state, "to" state, rates, and parameters lists.

```r
ctmc <- function()
  structure(list(from = NULL,
                 rate = NULL,
                 to = NULL,
                 params = NULL),
            class = "ctmc")
```

We want the syntax for constructing a CTMC to look like this:

```r
m <- ctmc() %>%
  add_edge(foo, a, bar) %>%
  add_edge(foo, 2*a, baz) %>%
  add_edge(foo, 4, qux) %>%
  add_edge(bar, b, baz) %>%
  add_edge(baz, a + x*b, qux) %>%
  add_edge(qux, a + UQ(x)*b, foo)
```

Therefore, we need to implement the add_edge such that it takes four arguments: the CTMC, the "from" state, the rate of the transition, and the "to" state. The CTMC is implicitly provided to the function calls when we are using the pipe operator. The other three arguments should be provided as expressions, and the add_edge function will implement a non-standard evaluation to handle them.

We want the "from" and "to" states to be single symbols, but we will translate these into strings that we can use as row and column names in the rate matrix for the CTMC. The rate associated with a transition should be an expression, and to get the scope of the expression right, we will translate it into a quosure. We will then extract the unbound variables in this expression—unbound in the environment in which the quosure is defined—and add them to the parameters of the model. The implementation looks like this:

```
add_edge <- function(ctmc, from, rate, to) {
  from <- enexpr(from); stopifnot(is_symbol(from))
  to <- enexpr(to); stopifnot(is_symbol(to))

  from <- as_string(from)
  to <- as_string(to)

  ctmc$from <- cons(from, ctmc$from)
  ctmc$to <- cons(to, ctmc$to)

  r <- enquo(rate)
  ctmc$rate <- cons(r, ctmc$rate)
  ctmc$params <- cons(collect_symbols_q(get_expr(r), get_env(r)),
                      ctmc$params)

  ctmc
}
```

We use enexpr for from and to since we want these symbols to be just that, symbols, and not something we will want to evaluate in any context. We use enquo for the rate parameter, on the other hand, because we do want to have its environment available when we evaluate the expression. We do not evaluate it yet, though. We cannot evaluate it until we know the parameters for the model, and we do not want those to be fixed inside the CTMC object. We use the rate environment, however, when extracting the unbound variables in the rate expression.

Generally, it is a good idea to be able to get some information about an object we construct by printing it, but the default print function for a ctmc object will show the list of lists. This representation, especially for the linked lists, can be hard to decipher. Instead, we can implement a print function for this class by defining a function with the name print.ctmc. The information we want to display is the parameters of the model and the edge structure, and we can implement this function like this:

```
print.ctmc <- function(x, ...) {
  from <- lst_to_list(x$from) %>% rev()
  to <- lst_to_list(x$to) %>% rev()
  rate <- lst_to_list(x$rate) %>% rev()
  parameters <- lst_to_list(x$params) %>%
    unlist() %>% unique() %>% rev()

  cat("CTMC:\n")
  cat("parameters:", paste(parameters), "\n")
  cat("transitions:\n")
  for (i in seq_along(from)) {
    cat(from[[i]], "->", to[[i]],
        "\t[", deparse(get_expr(rate[[i]])), "]\n")
  }
  cat("\n")
}
```

The implementation is straightforward. We translate the linked lists into lists to make them easier to work with when we loop over the edges, and we reverse them so we will display them in the order in which they were added to the model. With the linked lists, we prepend new edges, so they are represented in the opposite order than the one in which they were added. For the parameters, we remove duplications using unique as well. After that, we simply print the parameters as a list and print a line for each edge, showing the "from" and "to" states together with the rate expression on the edge. For the latter, we use get_expr to get the bare expression, rather than the quosure, and we use the function deparse to translate the expression into a string that we can print.

With the three functions we have defined so far, we can now create and print a continuous time Markov chain.

```
x <- 2
m <- ctmc() %>%
  add_edge(foo, a, bar) %>%
  add_edge(foo, 2*a, baz) %>%
  add_edge(foo, 4, qux) %>%
  add_edge(bar, b, baz) %>%
  add_edge(baz, a + x*b, qux) %>%
  add_edge(qux, a + UQ(x)*b, foo)
m
## CTMC:
## parameters: a b
## transitions:
## foo -> bar   [ a ]
## foo -> baz   [ 2 * a ]
## foo -> qux   [ 4 ]
## bar -> baz   [ b ]
## baz -> qux   [ a + x * b ]
## qux -> foo   [ a + 2 * b ]
```

This example shows that we can have expressions on the edges that are constants, such as the edge from foo to qux that has the rate four. We can have expressions with unbound variables, a, b, and 2*a. And we can have expressions that involve a bound variable, the last two edges. Notice here that the second-to-last edge, from baz to qux, has a rate expression that includes the (unevaluated) variable x, while the last edge, from qux to foo, contains the expression a + 2*b, where the value of x has been inserted. This is the difference between including a bound variable and unquoting it in the expression. Since x is a bound variable, it is not considered a parameter of the model, but in the second-to-last expression, it will be used when we evaluate the rate. If we change its value, we also change the value of the rate expression. For the last rate expression, we have already inserted the value of x, so here we will not change the expression by changing the value of x.

Constructing a Rate Matrix

We saw how we could translate a list of edges into a rate matrix in Chapter 3, but in this chapter, we want to do a little more. In Chapter 3, we had numeric rates on the edges; we now have expressions. Instead of translating the CTMC into a rate matrix, we will create a function for generating rate matrices—a function that, given values for the parameters of the Markov model, will provide us with the corresponding rate matrix.

We implement this functionality via a closure. We write a function that extracts the information we need to build the rate matrix from the ctmc object and then define a function for computing the rate matrix given the model's parameters. It then returns this closure function. The implementation can look like this:

```
get_rate_matrix_function <- function(ctmc) {
  from <- lst_to_list(ctmc$from) %>% rev()
  to <- lst_to_list(ctmc$to) %>% rev()
  rate <- lst_to_list(ctmc$rate) %>% rev()

  nodes <- c(from, to) %>% unique() %>% unlist()
  parameters <- lst_to_list(ctmc$params) %>%
    unlist() %>% unique() %>% rev()

  n <- length(nodes)
```

```
f <- function() {
  args <- as_list(environment())
  Q <- matrix(0, nrow = n, ncol = n)
  rownames(Q) <- colnames(Q) <- nodes
  for (i in seq_along(from)) {
    Q[from[[i]], to[[i]]] <- eval_tidy(rate[[i]], args)
  }
  diag(Q) <- -rowSums(Q)
  Q
}
formals(f) <- make_args_list(parameters)

f
}
```

Once again, we translate the linked lists into list objects and reverse them. We then get a list of unique nodes in the model by combining the from and to lists, removing duplicates, and we translate the resulting list into a vector that we will later use to set row and column names of the rate matrix. We extract the parameters for the model by translating the linked list into a list, and we then translate that into a vector, remove duplicates, and reverse the result to get the parameters in the order in which they were added to the edges.

The closure we define initially takes no formal arguments. We set those from the CTMC arguments after we have defined the function. We do it this way only because it is an easier way to define the function compared to constructing expressions and using something like the new_function construction we used earlier. Before we return the closure, it *will* have a list of formal arguments. Since we don't know what these will be, we use a trick to get hold of them inside the closure: we get the local environment before we define any local variables—so at this point it will contain only the parameters passed to the function call—and make a list out of them. That list, we can use later to over-scope the evaluation of the rate expressions inside the closure.

For the actual construction of the rate matrix, there is little to surprise. We get the size of the matrix from the number of states in the CTMC. We then create the matrix and name rows and columns according to the nodes they represent. Then we (tidy) evaluate all the rate expressions to fill in the cells of the matrix, and finally, we adjust the diagonal so all rows sum to zero.

We now have a command in our language for getting a rate matrix function.

```
Qf <- m %>% get_rate_matrix_function()
Qf

## function (a, b)
## {
##     args <- as_list(environment())
##     Q <- matrix(0, nrow = n, ncol = n)
##     rownames(Q) <- colnames(Q) <- nodes
##     for (i in seq_along(from)) {
##         Q[from[[i]], to[[i]]] <- eval_tidy(rate[[i]], args)
##     }
##     diag(Q) <- -rowSums(Q)
##     Q
## }
## <environment: 0x7fdf6e76fed8>
```

When we provide the model parameters to this function, we get the rate matrix.

```
Qf(a = 2, b = 4)

##      foo bar baz qux
## foo -10   2   4   4
## bar   0  -4   4   0
## baz   0   0 -10  10
## qux  10   0   0 -10
```

Remember that the edge from baz to qux holds an expression that refers to the global variable x. If we change the value of this variable, we also change the result of evaluating the Qf function.

```
x <- 1
Qf(a = 2, b = 4)

##       foo bar baz qux
## foo -10   2   4   4
## bar   0  -4   4   0
## baz   0   0  -6   6
## qux  10   0   0 -10
```

The edge from qux to foo, where we substitute the value for x at the time we created the edge, using UQ, does not change.

Traces

An *observation* for a continuous-time Markov chain is a *trace*—a sequence of states and at which time points we observe the states. In any real data analysis, we would probably write functions to obtain data from files, but since we are exploring domain-specific languages, let's write one for specifying traces. We will make traces depend on the CTMC that we want to use them with so we can test that the states in a trace are also states in the CTMC. If you want to use several CTMCs to analyze the same trace, you could remove these tests, or you could make the trace object depend on a list of legal states instead of a ctmc object.

We take the same approach as for the ctmc class: we write a function for creating an object to represent traces, and we then have functions for adding information to a trace. The information we want to store in a trace is a list of states and a list of time points in which we observe the states. For the consistency checks between CTMC and trace, we will also store the nodes in the CTMC. The constructor for the trace class looks like this:

```
ctmc_trace <- function(ctmc) {
  nodes <- c(lst_to_list(ctmc$from), lst_to_list(ctmc$to)) %>%
    unique %>% unlist
  structure(list(nodes = nodes, states = NULL, at = NULL),
            class = "ctmc_trace")
}
```

We add a verb to the language, a function adding observations of states at specific time points. This function mainly checks the consistency between states and the ctmc object and then adds states and time points to the ctmc_trace object's lists.

```
add_observation <- function(trace, state, at) {
  state <- enexpr(state)
  stopifnot(is_symbol(state))
  state <- as_string(state)
  stopifnot(state %in% trace$nodes)
  stopifnot(is.numeric(at))
  stopifnot(is.null(trace$at) || at > trace$at$car)

  trace$states <- cons(state, trace$states)
  trace$at <- cons(at, trace$at)

  trace
}
```

As for CTMC objects, we want a printing function for traces. Here, I will take a different approach than what we did for the ctmc print function. I will translate traces into data frames—tibble objects to be precise—and print the result. If I was writing a package for CTMCs, I might take a different approach, but I will use the transformation into data frames later to compute likelihoods, so I exploit the transformation in the print function as well. To translate a ctmc_trace object into a tibble object, we specialize the as_tibble function. After that, we specialize the print function.

```
as_tibble.ctmc_trace <- function(x, ...) {
  states <- x$states %>% lst_to_list() %>% unlist() %>% rev()
  at <- x$at %>% lst_to_list() %>% unlist() %>% rev()
  tibble(state = states, at = at)
}
print.ctmc_trace <- function(x, ...) {
  df <- as_tibble(x)
  cat("CTMC trace:\n")
  print(df)
}
```

We now have the functionality to create and print traces.

```
tr <- ctmc_trace(m) %>%
  add_observation(foo, at = 0.0) %>%
  add_observation(bar, at = 0.1) %>%
  add_observation(baz, at = 0.3) %>%
  add_observation(qux, at = 0.5) %>%
  add_observation(foo, at = 0.7) %>%
  add_observation(baz, at = 1.1)
tr

## CTMC trace:
## # A tibble: 6 x 2
##    state    at
##    <chr> <dbl>
## 1 foo   0.
## 2 bar   0.100
## 3 baz   0.300
```

```
## 4 qux    0.500
## 5 foo    0.700
## 6 baz    1.10
```

Computing Likelihoods

The final functionality we will implement for this example is for computing the likelihood of parameters given a CTMC and a trace. In Chapter 3, we saw how to translate a rate matrix into a transition-probability matrix by first multiplying the rate matrix by a scalar—the time period that has passed between two observations—and then (matrix-)exponentiating the result. We will reuse the function we implemented there.

```
transition_probabilities <- function(Q, t) expm(Q * t)
```

For computing the likelihood, we will create a verb in our domain-specific language that translates a CTMC and a trace into a function. This function will take the parameters of the CTMC as arguments—as the function for creating rate matrices we wrote earlier—and then return the likelihood for those parameters. This is a function we could then use for maximum-likelihood estimation by combining it with an optimization algorithm, of which there are several available in various R packages.

The implementation is straightforward. We get the rate matrix function from the ctmc object and translate the trace into a data frame and store the results in the closure of the function. Then we use the same trick as we used earlier to get the arguments inside the closure, evaluate the rate-matrix function to get the rate-matrix, and put the data from the data frame into a format we need for the computation. That computation is just running through the trace and computing the transition probabilities from two consecutive observations. Since it is a Markov model, the joint probability is the product of them. After we have created the closure, we set its formal arguments, similar to what we did with the rate-matrix function.

```
get_likelihood_function <- function(ctmc, trace) {
  rate_func <- ctmc %>% get_rate_matrix_function()
  trace_df <- as_tibble(trace)

  lhd_function <- function() {
    args <- as_list(environment())
    Q <- do.call(rate_func, args)

    n <- length(trace_df$state)
    from <- trace_df$state[-n]
    to <- trace_df$state[-1]
    delta_t <- trace_df$at[-1] - trace_df$at[-n]

    lhd <- 1
    for (i in seq_along(from)) {
      P <- transition_probabilities(Q, delta_t[i])
      lhd <- lhd * P[from[i],to[i]]
    }
    lhd
  }
  formals(lhd_function) <- formals(rate_func)

  lhd_function
}
```

That is it; we can now compute likelihoods for a CTMC.

```
lhd <- m %>% get_likelihood_function(tr)
lhd(a = 2, b = 4)

## [1] 0.00120108
```

In an actual data analysis context, we probably would want to compute the log likelihood instead. For traces of any useful length, the actual likelihood will lead to underflow since we are dealing with finite-bit floating-point numbers. Modifying the likelihood function to a log-likelihood function is a simple matter of changing the product to a sum and taking the log of `P[from[i],to[i]]`.

There might be more functionality you would like to add to a language like this, but even with the few functions we have implemented so far, we have a useful domain-specific language. We have not used any operator overloading to implement it; we didn't have to do that. We have used tidy evaluation extensively, though, to implement the non-standard evaluation we use for rate expressions.

CHAPTER 11

Pattern Matching

In languages such as ML or Haskell, you can define data types by specifying functions you will use to construct values of any given type. In itself, that is not that interesting, but combined with a pattern matching feature of these languages, you can write very succinct functions for transforming data structures.

In my book *Functional Data Structures in R* (Mailund, 2017a), I describe several algorithms that depend on the transformation of various trees based on their structure. Such transformations involve figuring out the current structure of a tree—does it have a left subtree? Is that tree a leaf? If it is a red-black search tree, what is the color of the tree? And the color of its right subtree? In the algorithms I presented in that book, most of the functions contained tens of lines of code just for matching such a tree structure.

With the language we implement in this chapter, we will make writing such transformation functions vastly more efficient. We will write two main constructions. The first is for defining a data structure, which we can use to define red-black search trees like this:

```
colour := R | B
rb_tree := E | T(col : colour, left : rb_tree, value, right :
rb_tree)
```

© Thomas Mailund 2018
T. Mailund, *Domain-Specific Languages in R*,
https://doi.org/10.1007/978-1-4842-3588-1_11

The second construction is used to match values of such types and then perform actions accordingly. A balancing function for red-black search trees can be implemented succinctly like this:

```
balance <- function(tree) {
  cases(tree,
        T(B,T(R,a,x,T(R,b,y,c)),z,d) -> T(R,T(B,a,x,b),y,
        T(B,c,z,d)),
        T(B,T(R,T(R,a,x,b),y,c),z,d) -> T(R,T(B,a,x,b),y,
        T(B,c,z,d)),
        T(B,a,x,T(R,b,y,T(R,c,z,d))) -> T(R,T(B,a,x,b),y,
        T(B,c,z,d)),
        T(B,a,x,T(R,T(R,b,y,c),z,d)) -> T(R,T(B,a,x,b),y,
        T(B,c,z,d)),
        otherwise -> tree)
}
```

This function is a mere eight lines, compared to the 42 lines of code used in *Functional Data Structures in R*, where I also use some pattern matching tricks but not a domain-specific language uniquely designed for it. Such a language is what we will implement in this chapter.

For the chapter, we will need to use the following packages:

```
library(rlang)
library(magrittr)
library(dplyr)
```

We will also need the make_args_list function we defined in Chapter 6.

```
make_args_list <- function(args) {
  res <- replicate(length(args), substitute())
  names(res) <- args
  as.pairlist(res)
}
```

For a package that implements the functionality described in this chapter, and more, see `https://mailund.github.io/pmatch/`.

Constructors

The key feature of this domain-specific language is the type constructors—how we define values of a given type. The pattern matching aspect of the DSL will consist of nested constructor calls, so it is how we define the constructors that is the essential aspect of the language.

Here, we are inspired by function calls. We will use a syntax for constructors that matches variables and function calls.

```
TYPEDEF ::= TYPENAME ':=' CONSTUCTORS
CONSTUCTORS ::= CONSTUCTOR | CONSTUCTOR '|' CONSTRUCTORS
CONSTRUCTOR ::= NAME | NAME '(' ARGS ')'
ARGS ::= ARG | ARG ',' ARGS
ARG ::= NAME | NAME : TYPE
TYPE ::= NAME
```

We define a new type by giving it a name, to the left of a := operator, and by putting a sequence of constructors on the right of the := operator. Constructors, then, are separated by | and are either single names or a name followed by arguments in parentheses, where an argument is either a single name or a name followed by : and then a type, where we require that a type is a name.

We implement this grammar by implementing the := operator. An assignment has the lowest precedence, which means that whatever we write to the left or right of this operator will be arguments to the function. We do not have to worry about an expression in our language being translated into some call object of a different type. We cannot override the other assignment operators, <-, ->, and =, so we have to use :=. Since this is also traditionally used to mean "defined to be equal to," it works quite well.

The approach we take in implementing this part of the pattern matching DSL is different from the examples we have seen earlier. We do not create a data structure that we can analyze nor do we evaluate expressions directly from expressions in our new language. Instead, we combine parsing expressions with code generation—we generate new functions and objects while we parse the specification. We add these functions, and other objects for constants, to the environment in which we call :=. Adding these objects to this environment allows us to use the constructors after we have defined them with no further coding, but it does mean that calling := will have side effects.

The construction function will expect a type name as its left-hand parameter and an expression describing the different ways of constructing elements of the type on its right-hand side. We will translate the left-hand side into a quosure because we want to get its associated environment. The right-hand side we will turn into an expression. For the construction specification, we do not want to evaluate any of the elements (unless the user invokes quasi-quotations). The left-hand side—the type we are defining—is just treated as a string since that is how the S3 system deals with types, so we will make sure it is a single symbol and then get the string representation of it. For this, we can use the quo_name function from rlang. The right-hand side we have to parse, but we delegate this to a separate function that we define next. Finally, we specify a function for pretty-printing elements of the new type we define.

```
`:=` <- function(data_type, constructors) {
  data_type <- enquo(data_type)
  constructors <- enexpr(constructors)

  stopifnot(quo_is_symbol(data_type))
  data_type_name <- quo_name(data_type)
  process_alternatives(
    constructors,
    data_type_name,
```

```
  get_env(data_type)
)

assign(paste0("toString.", data_type_name),
        deparse_construction, envir = get_env(data_type))
assign(paste0("print.", data_type_name),
        construction_printer, envir = get_env(data_type))
}
```

The last two statements in this function, the calls to assign, create functions for printing elements of the type we are creating. We will implement the deparse_construction and construction_printer functions in a moment. They extract information about values from meta-information we will store in objects of the new type, and we can use the same functions for all types we define in our language. We use them to specialize the toString and print functions for this specific type. The paste0 calls create the names of the specializations of the generic toString and print functions. The assign function then stores deparse_construction and construction_printer under the appropriate names in the environment we get from get_env(data_type), in other words, the environment where we define the type.

The expression on the right-hand side of := defines how we construct elements of the new type. We allow there to be more than one way to do this, and we separate the various choices using the or operator, |. This approach resembles how we describe different alternatives when we specify a grammar, so it is a natural choice. To process the right-hand side, we use the function process_alternatives.

```
process_alternatives <- function(constructors,
                                 data_type_name,
                                 env) {
  if (is_lang(constructors) && constructors[[1]] == "|") {
    process_alternatives(
```

```
      constructors[[2]],
      data_type_name,
      env
    )
    process_alternatives(
      constructors[[3]],
      data_type_name,
      env
    )
  } else {
    process_constructor(
      constructors,
      data_type_name,
      env
    )
  }
}
```

In addition to the constructor expression, we pass the name of the type and the environment we are defining it in as parameters. We do not use these directly in this function but merely pass them along. We will use them later when we create the actual constructors.

The process_alternatives function recursively parse the expression to get all alternatives separated by |. The actual constructors will be either a function or a symbol, so the constructor specifications will not have higher precedence than the or operator. The first time we see something that isn't a call to |, then, we have a constructor. We handle those using the process_constructor function.

```
process_constructor <- function(constructor,
                                data_type_name,
                                env) {
```

```r
if (is_lang(constructor)) {
  process_constructor_function(
    constructor,
    data_type_name,
    env
  )
} else {
  process_constructor_constant(
    constructor,
    data_type_name,
    env
  )
}
}
```

This function figures out whether what we are looking at is a function constructor or a constant, in other words, a symbol. We use the `is_lang` function to test whether we are looking at a function. It does the same as `is.call` from the `base` package; I just prefer the `rlang` functions for this chapter.

Constant constructors are the simplest. They are merely symbols, so to make them available for programmers, we need to define a value for each such symbol. We will use NA as the value of these variables and store some meta-information with them. We set the class, so the `construction_printer` function will be called when we try to print the object, and we set the attribute `constructor_constant` that we will later need for pattern matching.

```r
process_constructor_constant <- function(constructor,
                                         data_type_name,
                                         env) {
```

```
  stopifnot(is_symbol(constructor))
  constructor_name <- as_string(constructor)
  constructor_object <- structure(
    NA,
    constructor_constant = constructor_name,
    class = data_type_name
  )
  assign(constructor_name, constructor_object, envir = env)
}
```

For the function constructors, we need to create, you guessed it, functions. We analyze the arguments given to the constructor specification and build a function out of that, and this function we then store in the environment where the constructor is defined. We permit two kinds of parameters to a constructor: either a symbol or a symbol with a type. For the latter, we use the : operator. If a parameter is a : call, then we consider the left-hand side the parameter and the right-hand side the type. We use the types to guarantee that values we construct are of the expected kind. If there is no type specified, we will allow a parameter to hold any value. We use the following function to translate the list of parameters from a function constructor expression into a data frame where the first column holds the argument names and the second column holds their type. We use NA to indicate that we allow any type. The function works by first translating the arguments—that are in the form of a call object—into a list. We have to use the base as.list function for this, rather than the rlang as_list, since the latter will not translate call objects into lists. Once we have the arguments as a list, we map the process_arg function over the elements. This function creates a row for the data frame per element, and we combine the rows using the bind_rows function from dplyr.

```r
process_arguments <- function(constructor_arguments) {
  process_arg <- function(argument) {
    if (is_lang(argument)) {
      stopifnot(argument[[1]] == ":")
      arg <- quo_name(argument[[2]])
      type <- quo_name(argument[[3]])
      tibble::tibble(arg = arg, type = type)
    } else {
      arg <- quo_name(argument)
      tibble::tibble(arg = arg, type = NA)
    }
  }
  constructor_arguments %>%
    as.list %>%
    purrr::map(process_arg) %>%
    bind_rows
}
```

The process_constructor_function translates a function construction specification into a function. The first element of the specifications, which is a call object, is the name of the function. For the remaining elements, we translate them into a data frame using the function we just saw. After that, we need to create the function that will work as the constructor. Here, we create a closure without arguments and then add formal parameters afterward, as we did in the previous chapter, and we get the actual parameters that the closure is called with using the as_list(environment()) trick.

The value we return from the closure is just the list of arguments that are provided to it but tagged with a constructor attribute we can use for pattern matching and a class set to the type we are defining, something we use for type checking. The type checking is the chief part of the constructor. Here, we check that we get the right number of arguments and that they have the right type if a type was specified.

191

Once we have created the closure and set its formal arguments, we also update its class, so it is both a constructor and a function. Giving constructor functions the class constructor is also something we will need when pattern matching. Then we assign it to the environment associated with the specification to make it available to the programmer.

```
process_constructor_function <- function(constructor,
                                          data_type_name,
                                          env) {
  stopifnot(is_lang(constructor))
  constructor_name <- quo_name(constructor[[1]])
  constructor_arguments <- process_arguments(constructor[-1])

  # Create the constructor function
  constructor <- function() {
    args <- as_list(environment())

    # Type check!
    stopifnot(length(args) == length(constructor_arguments$arg))
    for (i in seq_along(args)) {
      arg <- args[[constructor_arguments$arg[i]]]
      type <- constructor_arguments$type[i]
      stopifnot(is_na(type) || inherits(arg, type))
    }

    structure(args,
              constructor = constructor_name,
              class = data_type_name)
  }
  formals(constructor) <- make_args_list(constructor_
arguments$arg)
```

```
  # Set meta information about the constructor
  class(constructor) <- c("constructor", "function")

  # Put the constructor in the binding scope
  assign(constructor_name, constructor, envir = env)
}
```

The only remaining function to write for the constructors is the function for printing them. Here, we write a function that translates a constructed object into a string; this function we can then use recursively to translate any constructed element into a string. We then just call this function in construction_printer, which is the function that is assigned to the specialized print function for any type we define.

There is nothing complicated in the function. We first check whether the object has an attribute constructor. Strictly speaking, only the function constructors have this—the constant constructors have the attribute constructor_constant, but the attire function will pick an attribute if it gets a unique prefix, so we also get that. If we do not have a constructor attribute, then it isn't an element constructed from something we have defined from our language, so it must be a value of some other type—we just convert it into a string and return this. We use the generic toString function for this. This function converts any object into a string. It's not necessarily a beautiful representation of the object, but you can specialize it if you need to do so.

If the object we have *is* a constructor, it is either a constant or the result of a constructor function call. If the latter, it will be a list. If it is a list, then we must convert all the elements in the list into strings and paste them together. Otherwise, the name of the constructor is the string representation of the object.

```
deparse_construction <- function(object) {
  constructor_name <- attr(object, "constructor")
  if (is_null(constructor_name)) {
```

```
    # This is not a constructor, so just get the value
    return(toString(object))
  }

  if (is_list(object)) {
    components <- names(object)
    values <- as_list(object) %>% purrr::map(deparse_
    construction)

    print_args <- vector("character", length = length(components))
    for (i in seq_along(components)) {
      print_args[i] <- paste0(components[i], "=", values[i])
    }
    print_args <- paste0(print_args, collapse = ", ")
    paste0(constructor_name, "(", print_args, ")")

  } else {
    constructor_name
  }
}
construction_printer <- function(x, ...) {
  cat(deparse_construction(x), "\n")
}
```

As an example of using the construction language, we can define a binary tree as either a tree with a left and right subtree or a leaf.

```
tree := T(left : tree, right : tree) | L(value : numeric)
```

We can use the constructors to create a tree:

```
x <- T(T(L(1),L(2)),L(3))
x

## T(left=T(left=L(value=1), right=L(value=2)),
right=L(value=3))
```

Values we create using these constructors can be accessed just as lists—which, in fact, they are—using the variable names we used in the type specification.

```
x$left$left$value
```

```
## [1] 1
```

```
x$left$right$value
```

```
## [1] 2
```

```
x$right$value
```

```
## [1] 3
```

The type checking is rather strict, however. We demand that the values we pass to the constructor functions are of the types we give in the specification—in the sense that they must inherit the class from the specification—and this can be a problem in some cases where R would otherwise ordinarily just convert values. In the specification for the L constructor, for example, we require that the argument is numeric. We will get an error if we give it an integer.

```
L(1L)
```

```
## Error: is_na(type) || inherits(arg, type) is not TRUE
```

This situation is where we can use the variant of parameters without a type:

```
tree := T(left : tree, right : tree) | L(value)
L(1L)
```

```
## L(value=1)
```

An alternative solution could be to specify more than one type in the specification. If you are interested, you can play with that. I will just leave it here and move on to pattern matching.

Pattern Matching

We want to implement pattern matching such that an expression like this:

```
cases(L(1),
      L(v) -> v,
      T(L(v), L(w)) -> v + w,
      otherwise -> 5)
```

```
## [1] 1
```

should return 1, since the pattern L(v) matches the value L(1) and we return v, which we expect to be bound to 1. Likewise, we want this expression to return 9 since v should be bound to 4 and w to 5 and we return the result of evaluating v + w.

```
cases(T(L(4), L(5)),
      L(v) -> v,
      T(L(v), L(w)) -> v + w,
      otherwise -> 5)
```

```
## [1] 9
```

We want the otherwise keyword to mean anything at all and use it as a default pattern, so in this expression, we want to return 5.

```
cases(T(L(1), T(L(4), L(5))),
      L(v) -> v,
      T(L(v), L(w)) -> v + w,
      otherwise -> 5)
```

```
## [1] 5
```

196

The syntax for pattern matching uses the right-arrow operator. This operator is usually an assignment. We cannot specialize arrow assignments, but we can still use them in a meta-programming function. We use an assignment operator for the same reasons as we had for using the := operator for defining types. Since assignment operators have the lowest precedence, we don't have to worry about how tight the operators to the left and right of the operator binds. We could also have used that operator here, but I like the arrow more for this function. It shows us what different patterns map to. You need to be careful with the -> operator, though, since it is syntactic sugar for <-. This means that once we have an expression that uses ->, we will actually see a call to <-, and the left- and right-hand sides will be switched.

The cases function will take a variable number of arguments. The first is the expression we match against, and the rest are captured by the three-dots operator. The expressions there should not be evaluated directly, so we capture them as quosures. We then iterate through them, split them into left-hand and right-hand sides, and test the left-hand side against the expression. The function we use for testing the pattern will return an environment that contains bound variables if it matches, and NULL otherwise. If we have a match, we evaluate the right-hand side in the quosure environment over-scoped by the environment we get from matching the pattern.

```
cases <- function(expr, ...) {
  matchings <- quos(...)

  for (i in seq_along(matchings)) {
    eval_env <- get_env(matchings[[i]])
    match_expr <- quo_expr(matchings[[i]])
    stopifnot(match_expr[[1]] == "<-")

    test_expr <- match_expr[[3]]
    result_expr <- match_expr[[2]]
```

```
    match <- test_pattern(expr, test_expr, eval_env)
    if (!is_null(match))
      return(eval_tidy(result_expr, data = match, env =
      eval_env))
  }

  stop("No matching pattern!")
}
```

In the test_pattern function we create the environment where we will bind matched variables. If the pattern is otherwise, we return the empty environment—no variables are bound there. Otherwise, we need to explore both pattern and expression recursively.

We use the function test_pattern_rec to do this, but we do not call it directly. Instead, we use a function called callCC. The name stands for *call with current continuation*, and it is a function that sometimes causes some confusion for people not intimately familiar with functional programming. There is no need for this confusion, however, because all the function does is provide us with a way to return to the point where we called callCC.

We wrap the test_pattern_rec function in a closure, tester, that is called with a function that we call escape. This is the function that callCC will provide. If, at any point, we call the function escape, it will terminate whatever we are doing and return to the point where we called callCC. This means we can use escape to get out of deep recursions if we find out at some point that the pattern doesn't match the expression. We do not need to propagate a failed match up the call stack through the recursive function calls. As soon as we call escape, we are taken back to the end of test_pattern. Whatever we called escape with—it is a function of a single parameter—will be the return value of the callCC call. So, if we find that a pattern doesn't match, we will call escape with NULL. This will then be the result of test_pattern. If we never call escape but instead return normally from test_pattern_rec, then what we return from that function will also be the return value of the `callCC` call. So, if we match the pattern and return an environment from `test_pattern_rec`, this will also be the return value of the `test_pattern` call.

```
test_pattern <- function(expr, test_expr, eval_env) {
  # Environment in which to store matched variables
  match_env <- env()

  if (test_expr == quote(otherwise))
    return(match_env)

  # Test pattern
  tester <- function(escape)
    test_pattern_rec(escape, expr, test_expr,
                     eval_env, match_env)
  callCC(tester)
}
```

It is in `test_pattern_rec` the real work is done. It analyzes the pattern expression, stored in the `test_expr` variable, and matches it against the value stored in the `expr` variable. It also takes two environments as parameters. One is the environment where the expression and pattern are defined; it needs this environment to look up variables to check what they are. The other is the environment in which it should bind variables from the pattern. It, of course, also knows the `escape` function that it can use if it finds out that the pattern isn't matching.

```
test_pattern_rec <- function(escape, expr, test_expr,
                             eval_env, match_env) {

  # Is this a function-constructor?
  if (is_lang(test_expr)) {
    func <- get(as_string(test_expr[[1]]), eval_env)
```

```r
  if (inherits(func, "constructor")) {
    # This is a constructor.
    # Check if it is the right kind
    constructor <- as_string(test_expr[[1]])
    expr_constructor <- attr(expr, "constructor")
    if (is_null(expr_constructor) ||
        constructor != expr_constructor)
      escape(NULL) # wrong type

    # Now check recursively
    for (i in seq_along(expr)) {
      test_pattern_rec(
        escape,
        expr[[i]], test_expr[[i+1]],
        eval_env, match_env
      )
    }

    # If we get here, the matching was successfull
    return(match_env)
  }
}

# Is this a constant-constructor?
if (is_symbol(test_expr) &&
    exists(as_string(test_expr), eval_env)) {
  constructor <- as_string(test_expr)
  val <- get(constructor, eval_env)
  val_constructor <- attr(val, "constructor_constant")
  if (!is_null(val_constructor)) {
    expr_constructor <- attr(expr, "constructor")
    if (is_null(expr) || constructor != expr_constructor)
      escape(NULL) # wrong type
```

```
    else
        return(match_env) # Successfull match
    }
}

# Not a constructor.
# Must be a value to compare with or a variable to bind to
if (is_symbol(test_expr)) {
    assign(as_string(test_expr), expr, match_env)
} else {
    value <- eval_tidy(test_expr, eval_env)
    if (expr != value) escape(NULL)
}

match_env
}
```

There are three cases to consider. The test_expr is a constructor function, a constructor constant, or something else.

If test_expr is a function call, we test this using is_lang from rlang, then it might be a function constructor. To figure out whether it is, we look the function name up in the evaluation environment, in other words, the environment where the test pattern was written. We then test whether the function inherits a constructor. The functions we create in our DSL will also be constructor objects, so if it does inherit constructor, we know we have such one. We then check whether expr has an attribute constructor. If it was generated by a call to a constructor, it will. If it doesn't, then we cannot have a match, and we escape with NULL. We also escape if the names of the constructors do not match. If they do, we iterate through all the elements in the pattern and expression calls and attempt to match these. If they do not match, we will never return from a recursive call—they will have used the escape function to jump directly to the callCC point in test_pattern. If they return, the pattern did match the expression, and we return the match_env that now contains any variables that were bound in the matching.

If test_expr is not a function call, it might still be a constructor. If it is, then it will be a symbol, and the symbol will be a variable in the eval_env scope—we test this with the exists function. These two tests tell us only that there is a variable with the name from test_expr in eval_env. Not that it is a constant constructor. To test this, we get the value the variable evaluates to, using the get function, and check if it has an attribute called constructor_constant. If it does, then that is the name of the constructor, and we can test that against the value in expr that either will be the object that represents the constant (in which case we have a match) or will be something else (in which case we escape).

If we get past the first two tests, we do not have a constructor. We now either have a value that we must test against expr or have a variable that we should bind to the value of expr. Here, I have decided that any symbol will be interpreted as a variable that should be bound, and anything else should be a value that we evaluate and compare against expr. We could also have checked if the variable was bound and used its value in that case, but that could clearly lead to hard-to-fix bugs. In any case, you can always use quasi-quoting to achieve the same effect.

```
x <- 1
y <- 2
cases(L(1),
      L(!!x) -> "x",
      L(!!y) -> "y")

## [1] "x"
```

So, we bind any variable by assigning the value of expr to the symbol in the match_env environment. Anything that isn't a symbol should be a value that is the same as expr. To get this value, we evaluate the test_expr in the eval_env.

That was the entire implementation of pattern matching. It might not be trivial code, but it is not horribly complicated either, and we have created an efficient language in less than 200 lines of code.

We can try it, now, by implementing a depth-first traversal of the binary tree type we defined earlier. The following function is a simple traversal that adds together all the values found in leaves. The match call considers the basis case (a leaf) and the recursive case (a tree with a left and right subtree) and doesn't need an otherwise case.

```
dft <- function(tree) {
  cases(tree,
        L(v) -> v,
        T(left, right) -> dft(left) + dft(right))
}

dft(L(1))

## [1] 1

dft(T(L(1),L(2)))

## [1] 3

dft(T(T(L(1),L(2)),L(3)))

## [1] 6
```

Lists

We have used linked lists many places in this book, but the functions we have used had to access the list elements in a list. We can define linked lists using our new language like this:

```
linked_list := NIL | CONS(car, cdr : linked_list)
```

A list either is empty (we use the constant NIL to represent that) or has a head element (`car`) and a tail (`cdr`). Since we implement values we construct from our language as `list` objects, we automatically get the linked-list implementation from this specification.

With pattern matching, we can write simple functions for manipulating lists. For example, the following function reverses a list using an accumulator list. In the base case, when the first list is empty, we return the accumulator. Otherwise, we take the head of the list and prepend it to the accumulated list and then recurse. We force evaluation of the accumulator to avoid lazy evaluation. With lazy evaluation, we would be building larger and later CONS expressions that would not be evaluated until the end of the recursion. At this point, we might have built the expression too large for the stack space needed to evaluate the functions. See, for example, Mailund (2017b) and Mailund (2017a) for explanations of how recursion and lazy evaluation can collide to exceed the stack space.

```
reverse_list <- function(lst, acc = NIL) {
  force(acc)
  cases(lst,
        NIL -> acc,
        CONS(car, cdr) -> reverse_list(cdr, CONS(car, acc)))
}
```

We can write a similar function to compute the length of a list. This will follow the same pattern of using an accumulator, which we return in the base case and update in the recursive case.

```
list_length <- function(lst, acc = 0) {
  force(acc)
  cases(lst,
        NIL -> acc,
        CONS(car, cdr) -> list_length(cdr, acc + 1))
}
```

A function for translating a linked list into a `list` object is a little more involved but can still be written succinctly using pattern matching. We need to figure out the length of the `list` object first, then allocate it, and finally iterate through the linked list to update the `list`. We use a closure, f, to recursively traverse the linked list. In the base case, I return NULL. It doesn't matter what we return here since we only do the recursion for its side effects, which are handled in the recursive case. Here, I use a code block—the curly braces operator—to evaluate two statements. The first updates the list v, and the second continues the recursion. It is precisely because we use tidy evaluation that this function works. It is essential that the assignment we do in the recursive case is evaluated in the environment where we write the expression. Otherwise, we would not be updating the correct list.

```
list_to_vector <- function(lst) {
  n <- list_length(lst)
  v <- vector("list", length = n)
  f <- function(lst, i) {
    force(i)
    cases(lst,
          NIL -> NULL,
          CONS(car, cdr) -> {
            v[[i]] <<- car
            f(cdr, i + 1)
            }
          )
  }
  f(lst, 1)
  v %>% unlist
}
```

Translating a list to a linked list is simple enough and doesn't use any pattern matching—we cannot pattern match on list objects after all.

```
vector_to_list <- function(vec) {
  lst <- NIL
  for (i in seq_along(vec)) {
    lst <- CONS(vec[[i]], lst)
  }
  reverse_list(lst)
}
```

With these few functions, we can translate to and from vectors and work with lists.

```
lst <- vector_to_list(1:5)
list_length(lst)
```

```
## [1] 5
```

```
list_to_vector(lst)
```

```
## [1] 1 2 3 4 5
```

```
lst %>% reverse_list %>% list_to_vector
```

```
## [1] 5 4 3 2 1
```

Extending the functionality of linked lists with additional functions, I will leave as an exercise to the interested reader. You can experiment to your heart's desire.

Search Trees

As another example, we can implement search trees. These are trees, containing ordered values, that satisfy the recursive property that all elements in the left subtree of a search tree will have values less than the value stored at the root of the tree, and all elements in the right subtree of a search tree will have values larger than the value in the root.

To define search trees, we try a different approach than the binary trees from earlier. We define an empty tree, E, and a tree with two subtrees, left and right, and a value.

```
search_tree :=
  E | T(left : search_tree, value, right : search_tree)
```

We will just implement two functions for the example, insertion and test for membership. For more functions on search tree, I will refer to Mailund (2017a, Chapter 6).

Both functions search recursively down the tree until they either find the value they want to insert or the value we want to check membership for is in the tree, respectively. For insertion, the base case, when it hits a leaf, is to insert the element there. It will create a tree, with two empty subtrees, containing the value. The recursive function builds a tree in the recursion by using the T constructor to create new trees in each recursive call. Thus, the tree created at a leaf will be put into the updated tree that the insert function creates. For the member function, we do not need to update the tree. If we hit a leaf, we know that the element is not in the tree, and we can just return FALSE. In both functions, the search checks the value in the tree they see in the recursive case. If the value there is greater than x, then the only place x could be found would be in the left subtree, so we continue the search there. If, on the other hand, x is greater than the value, then we must search in the right subtree. If it is neither smaller than nor greater than the value, it must be equal to the value. For insertion, this means we do not have to do anything, and we can just return the tree that already contains x. For the membership test, we can return TRUE.

```
insert <- function(tree, x) {
  cases(tree,
        E -> T(E, x, E),
        T(left, val, right) -> {
          if (x < val)
            T(insert(left, x), val, right)
          else if (x > val)
            T(left, val, insert(right, x))
          else
            T(left, x, right)
        })
}

member <- function(tree, x) {
  cases(tree,
        E -> FALSE,
        T(left, val, right) -> {
          if (x < val) member(left, x)
          else if (x > val) member(right, x)
          else TRUE
        })
}
```

We can build a tree like this:

```
tree <- E
for (i in sample(2:4))
  tree <- insert(tree, i)
```

Once the tree is built, we can test membership like this:

```
for (i in 1:6) {
  cat(i, " : ", member(tree, i), "\n")
}
```

```
## 1  :  FALSE
## 2  :  TRUE
## 3  :  TRUE
## 4  :  TRUE
## 5  :  FALSE
## 6  :  FALSE
```

The worst-case time usage for both of these functions is proportional to the depth of the tree, and that can be linear in the number of elements stored in the tree. If we keep the tree balanced, though, the time is reduced to logarithmic in the size of the tree. A classical data structure for keeping search trees balanced is so-called red-black search trees. Implementing these using pointer or reference manipulation in languages such as C/C++ or Java can be quite challenging, but in a functional language, balancing such trees is a simple matter of transforming trees based on local structure. See, for example, Okasaki (1999), Germane and Might (2014), or Mailund (2017a).

Red-black search trees are binary search trees where each tree has a color associated, either red or black. We can define colors using constructors like this:

```
colour :=
  R | B
```

We add a color to all nonempty trees like this:

```
rb_tree :=
  E | T(col : colour, left : rb_tree, value, right : rb_tree)
```

Except for including the color in the pattern matching, the member function for this data structure is the same as for the plain search tree.

```
member <- function(tree, x) {
  cases(tree,
        E -> FALSE,
        T(col, left, val, right) -> {
```

```
            if (x < val) member(left, x)
            else if (x > val) member(right, x)
            else TRUE
        })
}
tree <- T(R, E, 2, T(B, E, 5, E))
for (i in 1:6) {
  cat(i, " : ", member(tree, i), "\n")
}
```

```
## 1  :  FALSE
## 2  :  TRUE
## 3  :  FALSE
## 4  :  FALSE
## 5  :  TRUE
## 6  :  FALSE
```

What keeps red-black search trees balanced is that we always enforce these two invariants:

- No red node has a red parent.

- Every path from the root to a leaf has the same number of black nodes.

If every path from root to a leaf has the same number of black nodes, then the tree is perfectly balanced if we ignored the red nodes. Since no red node has a red parent, the longest path, when red nodes are considered, can be no longer than twice the length of the shortest path.

These invariants can be guaranteed by always inserting new values in red leaves, potentially invalidating the first invariant, and then rebalancing all subtrees that invalidate this invariant and at the end setting the root to be black. The rebalancing is done when returning from the recursive insertion calls that otherwise work as insertion in the plain search tree.

```
insert_rec <- function(tree, x) {
  cases(tree,
        E -> T(R, E, x, E),
        T(col, left, val, right) -> {
          if (x < val)
            balance(T(col, insert_rec(left, x), val, right))
          else if (x > val)
            balance(T(col, left, val, insert_rec(right, x)))
          else
            T(col, left, x, right) # already here
        })
}
insert <- function(tree, x) {
  tree <- insert_rec(tree, x)
  tree$col <- B
  tree
}
```

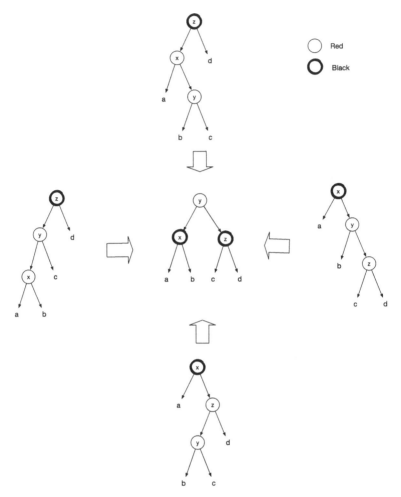

Figure 11-1. *Rebalancing transformations when inserting into a red-black search tree*

Figure 11-1 shows the transformation rules for the balance function. Whenever we see any of the four trees on the edges, we have to transform it into the one in the middle. The implementation I presented in Mailund (2017a) contained mostly code for testing the structure of the tree to match and very little to construct the modified tree. With pattern matching, we can implement these rules by matching for each of the four cases like this:

```
balance <- function(tree) {
  cases(tree,
        T(B,T(R,a,x,T(R,b,y,c)),z,d) -> T(R,T(B,a,x,b),y,
        T(B,c,z,d)),
        T(B,T(R,T(R,a,x,b),y,c),z,d) -> T(R,T(B,a,x,b),y,
        T(B,c,z,d)),
        T(B,a,x,T(R,b,y,T(R,c,z,d))) -> T(R,T(B,a,x,b),y,
        T(B,c,z,d)),
        T(B,a,x,T(R,T(R,b,y,c),z,d)) -> T(R,T(B,a,x,b),y,
        T(B,c,z,d)),
        otherwise -> tree)
}
```

This is the function we used to motivate the domain-specific language, and so we come full circle, having implemented the language we wanted.

CHAPTER 12

Dynamic Programming

As yet another example, we will create a domain-specific language for specifying dynamic programming algorithms. Dynamic programming is a technique used for speeding up recursive computations. When we need to compute a quantity from a recursion that splits into other recursions and these recursions overlap—so the computation involves evaluating the same recursive calls multiple times—we can speed up the computation by memorizing the results of the recursions. If we *a priori* know the values we will need to compute, we can build up a table of these values from the basic cases up to the final value instead of from the top-level recursion and down where we need more bookkeeping to memorize results.

Take a classical example such as *Fibonacci numbers*. The Fibonacci number $F(n) = 1$ if n is 1 or 2; otherwise, $F(n) = F(n-1) + F(n-2)$. To compute $F(n)$, we need to recursively compute $F(n-1)$ and $F(n-2)$. To compute $F(n-1)$, we need to compute $F(n-2)$ and $F(n-3)$, which obviously overlap the recursions needed to compute $F(n-2)$, which is computed from $F(n-3)$ and $F(n-4)$.

We could compute the n'th Fibonacci number recursively, but we would have to compute an exponential number of recursions. Instead, we could memorize the results of $F(m)$ for each m we use in the recursions to avoid recomputing values. A much more straightforward approach is to compute $F(m)$ for $m = 1,\ldots,n$ in that order and store the results in a table.

© Thomas Mailund 2018
T. Mailund, *Domain-Specific Languages in R*,
https://doi.org/10.1007/978-1-4842-3588-1_12

```
n <- 10
F <- vector("numeric", length = n)
F[1] <- F[2] <- 1
for (m in 3:n) {
    F[m] <- F[m-1] + F[m-2]
}
F[n]
```

[1] 55

Another, equally classical, example is computing the *edit distance* between two strings, x and y, of length n and m, respectively. This is the minimum number of transformations—character substitutions, deletions, or insertions—needed to translate x into y and can be defined recursively for $i=1,\ldots,n+1$ and $j=1,\ldots,m+1$ as follows:

$$
E[i,j] = \begin{cases} & i-1 & j=1 \\ & j-1 & i=1 \\ \min \begin{cases} E[i-1,j]+1, & \text{(Insertion)} \\ E[i,j-1]+1, & \text{(Deletion)} \\ E[i-1,j-1]+1_{x[i-1] \neq y[j-1]} & \text{(Substitution)} \end{cases} \end{cases}
$$

There are two base cases, capturing the edit distance of a prefix of x against an empty string or an empty string against a prefix of y, and then there are three cases for the recursion: one for insertion, where we move from $x=[1,\ldots,i-2]$ to $x[1,\ldots,i-1]$ against $y[1,\ldots,j-1]$; one for deletion, where we move from $y[1,\ldots,j-2]$ to $y[1,\ldots,j-1]$ against $x[1,\ldots,i-1]$; and finally substitution, where we move from $x=[1,\ldots,i-2]$ against $y[1,\ldots,j-2]$ to $x=[1,\ldots,i-2]$ against $y[1,\ldots,j-2]$. The cost of this is 0 if $x[i-1]=y[j-1]$ and 1 if $x[i-1] \neq y[j-1]$, which is captured by the indicator variable $1_{x[i-1] \neq y[j-1]}$.

This recursion is also readily translated into a dynamic programming algorithm. For test purposes, we can construct these two sequences:

```
x <- c("a", "b", "c")
y <- c("a", "b", "b", "c")
```

where we can go from x to y by inserting one b, so the edit distance is 1.

Computing the recursion, using dynamic programming, could look like this:

```
n <- length(x)
m <- length(y)
E <- vector("numeric", length = (n + 1) * (m + 1))
dim(E) <- c(n + 1, m + 1)
for (i in 1:(n + 1))
    E[i, 1] <- i - 1
for (j in 1:(m + 1))
    E[1, j] <- j - 1
for (i in 2:(n + 1)) {
    for (j in 2:(m + 1)) {
        E[i, j] <- min(
            E[i - 1, j] + 1,
            E[i, j - 1] + 1,
            E[i - 1, j - 1] + (x[i - 1] != y[j - 1])
        )
    }
}
E
```

```
##      [,1] [,2] [,3] [,4] [,5]
## [1,]    0    1    2    3    4
## [2,]    1    0    1    2    3
## [3,]    2    1    0    1    2
## [4,]    3    2    1    1    1
```

The edit distance can be obtained by the bottom-right cell in this table:

```
E[n + 1, m + 1]
```

```
## [1] 1
```

For both of these examples, it is straightforward to translate the recursions into dynamic programming algorithms, but the declarations of the problems—expressed in the recursions—are lost in the implementations, which are the loops where we fill out the tables.

We want to construct a domain-specific language that lets us specify a dynamic programming algorithm from a recursion and lets the language build the loops and computations for us. For example, we should be able to specify a recursion like this:

```
fib <- {
    F[n] <- 1 ? n <= 2
    F[n] <- F[n - 1] + F[n - 2]
} %where% {
    n <- 1:10
}
```

and have this expression build the Fibonacci table:

```
fib
```

```
## [1]  1  1  2  3  5  8 13 21 34 55
```

In the recursion, we use the ? operator to specify the case in which a given rule should be used. You are probably familiar with ? used to get documentation for functions, but the R parser also considers it an infix operator and one with the lowest precedence at all—lower than even <- assignment. This, as it turns out, is convenient for this DSL. We specify the recursion cases as assignments, and by having an operator with lower precedence than an assignment, we will always have a ? call at the top level of the call expression if the rule has a condition associated with it.

Similarly to the Fibonacci recursion, we should be able to specify the edit-distance computation like this:

```
x <- c("a", "b", "c")
y <- c("a", "b", "b", "c")
edit <- {
  E[1,j] <- j - 1
  E[i,1] <- i - 1
  E[i,j] <- min(
      E[i - 1,j] + 1,
      E[i,j - 1] + 1,
      E[i - 1,j - 1] + (x[i - 1] != y[j - 1])
  ) ? i > 1 && j > 1
} %where% {
    i <- 1:(length(x) + 1)
    j <- 1:(length(y) + 1)
}
```

and get the table computed:

```
edit
```

```
##      [,1] [,2] [,3] [,4] [,5]
## [1,]    0    1    2    3    4
## [2,]    1    0    1    2    3
## [3,]    2    1    0    1    2
## [4,]    3    2    1    1    1
```

Here, the first two cases are valid only when I or j is 1, respectively. We could specify this using ?, but I'm taking another approach. I use the index pattern on the left-hand side of the assignments to specify that an index variable should match a constant. This gives us semantics similar to the pattern matching in the previous chapter.

We could also have used this syntax for the Fibonacci recursion, like this:

```
{
    F[1] <- 1
    F[2] <- 1
    F[n] <- F[n - 1] + F[n - 2]
} %where% {
    n <- 1:10
}
```

```
## [1]  1  1  2  3  5  8 13 21 34 55
```

For the semantics of evaluating the recursion, I will use the first rule where both index patterns and ? conditions are satisfied to compute a value. Since we always pick the first such expression, we don't need to explicitly specify the conditions for the general case in the edit-distance specification, for example.

```
{
  E[1,j] <- j - 1
  E[i,1] <- i - 1
  E[i,j] <- min(
      E[i - 1,j] + 1,
      E[i,j - 1] + 1,
      E[i - 1,j - 1] + (x[i - 1] != y[j - 1])
  )
} %where% {
    i <- 1:(length(x) + 1)
    j <- 1:(length(y) + 1)
}
```

```
##        [,1] [,2] [,3] [,4] [,5]
## [1,]    0    1    2    3    4
## [2,]    1    0    1    2    3
## [3,]    2    1    0    1    2
## [4,]    3    2    1    1    1
```

Parsing Expressions

The examples give us an idea about what grammar we want for the
language. We want it to look somewhat like this:

```
DYNPROG_EXPR ::= RECURSIONS '%where%' RANGES
RECURSIONS ::= '{' PATTERN_ASSIGNMENTS '}'
RANGES ::= '{' RANGES_ASSIGNMENTS '}'
```

At the top level, the language is implemented using the user-defined
infix operator %where%. On the left-hand side of the operator, we want
a specification of the recursion, and on the right-hand side, we want
a specification of the ranges the algorithm should loop over. We can
implement the operator like this:

```
`%where%` <- function(recursion, ranges) {
    parsed <- list(
        recursions = parse_recursion(rlang::enquo(recursion)),
        ranges = parse_ranges(rlang::enquo(ranges))
    )
    eval_dynprog(parsed)
}
```

We simply parse the left-hand and right-hand sides and then call the
function eval_dynprog to run the dynamic programming algorithm and
return the result.

The simplest aspect of the language is the specification of the ranges. Here, we define variables and associate them with sequences they should iterate over.

```
RANGES_ASSIGNMENTS ::= RANGES_ASSIGNMENT
                    |  RANGES_ASSIGNMENT ';' RANGES_ASSIGNMENTS
```

We want one or more assignments. Here, I've specified that we separate them by ;, but we will also accept newlines; we simply use what R uses to separate statements.

Assignments take the following form:

```
RANGES_ASSIGNMENT ::= RANGE_INDEX '<-' RANGE_EXPRESSION
```

I won't break this down further, but just specify that RANGE_INDEX should be an R variable and RANGE_EXPRESSION an R expression that evaluates to a sequence.

In the %where% operator, we translate the ranges specification into a quosure, so we know in which scope to evaluate the values for the ranges, and then we process the result like this:

```r
parse_ranges <- function(ranges) {
    ranges_expr <- rlang::get_expr(ranges)
    ranges_env <- rlang::get_env(ranges)

    stopifnot(ranges_expr[[1]] == "{")
    ranges_definitions <- ranges_expr[-1]

    n <- length(ranges_definitions)
    result <- vector("list", length = n)
    indices <- vector("character", length = n)

    for (i in seq_along(ranges_definitions)) {
        assignment <- ranges_definitions[[i]]
```

```
        stopifnot(assignment[[1]] == "<-")
        range_var <- as.character(assignment[[2]])
        range_value <- eval(assignment[[3]], ranges_env)

        indices[[i]] <- range_var
        result[[i]] <- range_value
    }

    names(result) <- indices
    result
}
```

First, we extract the expression and the environment of the quosure. We will process the former and use the latter to evaluate expressions. We expect the ranges to be inside curly braces, so we test this and extract the actual specifications. We iterate over them, expecting each to be an assignment, where we can get the index variable on the left-hand side and the expression for the actual ranges on the right-hand side. We evaluate the expressions, in the quosure scope, and build a list as the parse result. This list will contain one item per range expression. The name of the item will be the iterator value, and the value will be the result of evaluating the corresponding expression.

```
parse_ranges(rlang::quo({
    n <- 1:10
}))
```

```
## $n
##  [1]  1  2  3  4  5  6  7  8  9 10
```

```
parse_ranges(rlang::quo({
    i <- 1:(length(x) + 1)
    j <- 1:(length(y) + 1)
}))
```

```
## $i
## [1] 1 2 3 4
##
## $j
## [1] 1 2 3 4 5
```

The recursion specifications are slightly more complicated.

```
PATTERN_ASSIGNMENTS ::= PATTERN_ASSIGNMENT
                      | PATTERN_ASSIGNMENT ';' PATTERN_
                                                ASSIGNMENTS
```

where

```
PATTERN_ASSIGNMENT ::= PATTERN '<-' RECURSION
                     | PATTERN '<-' RECURSION '?' CONDITION
PATTERN ::= TABLE '[' INDICES ']'
```

We won't break the grammar down further than this. Here TABLE
is a variable, INDICES is a comma-separated sequence of variables or
expressions, and both RECURSION and CONDITION are R expressions.

With this grammar, we need to extract three different pieces of
information: the index patterns, so we can match against that; the ?
conditions, such that we can check those; and finally the actual recursions.
The parser doesn't have to be much more complicated than for the ranges,
though, as long as we just collect these three properties of each recursive
case and put them in lists.

```
parse_recursion <- function(recursion) {
    recursion_expr <- rlang::get_expr(recursion)
    recursion_env <- rlang::get_env(recursion)

    stopifnot(recursion_expr[[1]] == "{")
    recursion_cases <- recursion_expr[-1]
```

```
n <- length(recursion_cases)
patterns <- vector("list", length = n)
conditions <- vector("list", length = n)
recursions <- vector("list", length = n)

for (i in seq_along(recursion_cases)) {
    case <- recursion_cases[[i]]

    condition <- TRUE
    stopifnot(rlang::is_call(case))
    if (case[[1]] == "?") {
        # NB: The order matters here!
        condition <- case[[3]]
        case <- case[[2]]
    }

    stopifnot(case[[1]] == "<-")
    pattern <- case[[2]]
    recursion <- case[[3]]

    patterns[[i]] <- pattern
    recursions[[i]] <- recursion
    conditions[[i]] <- condition
}

list(
    recursion_env = recursion_env,
    patterns = patterns,
    conditions = conditions,
    recursions = recursions
)
}
```

The function is not substantially different from the parser for the ranges. We extract the quosure environment and the expression, check that it is a call to curly braces, and then loop through the cases.

If a case is a call to ?, we know it has a condition. Since ? has the lowest precedence of all operators, it *will* be the top-level call if a condition exists—unless, at least, users get cute with parentheses, but then they get what they deserve. As a default condition, we use TRUE. This way, we don't have to deal with special cases when there is no ? condition specified; but if there is, we replace TRUE with the expression. Otherwise, we just collect all the information for each recursive case.

We do not evaluate the recursions in the parser, so we do not use the quosure environment in the parser. Instead, we return it together with the parsed information. We will need it later when we evaluate the recursion expressions.

We can test the parser on the Fibonacci recursion to see how it behaves.

```
parse_recursion(rlang::quo({
  F[n] <- n * F[n - 1] ? n > 1
  F[n] <- 1            ? n <= 1
}))

## $recursion_env
## <environment: R_GlobalEnv>
##
## $patterns
## $patterns[[1]]
## F[n]
##
## $patterns[[2]]
## F[n]
##
##
```

```
## $conditions
## $conditions[[1]]
## n > 1
##
## $conditions[[2]]
## n <= 1
##
##
## $recursions
## $recursions[[1]]
## n * F[n - 1]
##
## $recursions[[2]]
## [1] 1
```

Similarly, we can use it to parse the edit-distance recursions.

```
parse_recursion(rlang::quo({
  E[1,j] <- j - 1
  E[i,1] <- i - 1
  E[i,j] <- min(
      E[i - 1,j] + 1,
      E[i,j - 1] + 1,
      E[i - 1,j - 1] + (x[i - 1] != y[j - 1])
  ) ? i > 1 && j > 1
}))

## $recursion_env
## <environment: R_GlobalEnv>
##
## $patterns
## $patterns[[1]]
## E[1, j]
```

```
##
## $patterns[[2]]
## E[i, 1]
##
## $patterns[[3]]
## E[i, j]
##
##
## $conditions
## $conditions[[1]]
## [1] TRUE
##
## $conditions[[2]]
## [1] TRUE
##
## $conditions[[3]]
## i > 1 && j > 1
##
##
## $recursions
## $recursions[[1]]
## j - 1
##
## $recursions[[2]]
## i - 1
##
## $recursions[[3]]
## min(E[i - 1, j] + 1, E[i, j - 1] + 1, E[i - 1, j - 1] + (x[i -
##      1] != y[j - 1]))
```

Evaluating Expressions

To evaluate a dynamic programming expression, we need to iterate over all the ranges, compute the first value associated with satisfied conditions and patterns, and put the result in the dynamic programming table. We *could* construct expressions with nested loops for the ranges to do this, but there is a simpler way. We can construct all combinations of ranges using the expression do.call(expand.grid, ranges). For the Fibonacci example, we get this:

```
ranges <- parse_ranges(rlang::quo({
    n <- 1:10
}))
head(do.call(expand.grid, ranges))
```

```
##    n
## 1  1
## 2  2
## 3  3
## 4  4
## 5  5
## 6  6
```

Except that we have a name associated with the range, this is just the input. For the edit-distance example, we see the full use of this expression here:

```
ranges <- parse_ranges(rlang::quo({
    i <- 1:(length(x) + 1)
    j <- 1:(length(y) + 1)
}))
head(do.call(expand.grid, ranges))
```

```
##    i j
## 1 1 1
## 2 2 1
## 3 3 1
## 4 4 1
## 5 1 2
## 6 2 2
```

Here, we get all combinations of *i* and *j* from the ranges. Instead of constructing expressions with nested loops, we can iterate through the rows of the table we get from this invocation. If we extract a row, we get a vector with named range indices.

```
edit_ranges <- do.call(expand.grid, ranges)
edit_ranges[1,]
```

```
##    i j
## 1 1 1
```

```
edit_ranges[2,]
```

```
##    i j
## 2 2 1
```

Using the with function, we can exploit this to create an environment where range indices are bound to values, and it is in precisely such environments we want to evaluate the recursions.

```
with(edit_ranges[3,],
    cat(i, j, "\n")
)
```

```
## 3 1
```

Looping over the rows in a table we create this way will be the core of our algorithm. Before we get to implementing this, however, we need to construct the expressions to insert into the body of the loop.

We start with creating code for testing patterns against range-index variables. Here, we will keep the syntax simpler than the pattern matching from the previous chapter. We assume that the indices in the left-hand side of recursions consist of index variables or constants, and we build a test that simply compares the index-variable names, which we get from the ranges parser, against the values in the recursion specification. We can construct the function as simply as this:

```
make_pattern_match <- function(pattern, range_vars) {
    matches <- vector("list", length = length(range_vars))
    stopifnot(pattern[[1]] == "[")
    for (i in seq_along(matches)) {
        matches[[i]] <- call(
            "==",
            pattern[[i + 2]],
            range_vars[[i]]
        )
    }
    rlang::expr(all(!!! matches))
}
```

We construct a list of == comparisons between the index patterns used in the specification and the values the range indices can iterate over. For the latter, we first need to translate strings into symbols, so we are sure to compare the *values* of pattern variables and not check that constants match the name of the variables. We do not perform that conversion inside this function; we will expect that the caller has already done this. The reason that we use index i + 2 for the patterns is that patterns are in the form of a call to the [operator, so index 1 is the [symbol and index 2 is the table name. The indices follow after those two argument-elements.

A couple of examples should illustrate how the pattern-matching predicates are constructed.

```
make_pattern_match(rlang::expr(F[1]), list(as.symbol("n")))
```

```
## all(1 == n)
```

```
make_pattern_match(rlang::expr(F[n]), list(as.symbol("n")))
```

```
## all(n == n)
```

```
make_pattern_match(rlang::expr(E[1,j]),
                   list(as.symbol("i"), as.symbol("j")))
```

```
## all(1 == i, j == j)
```

We test all range variables, even when they are completely free to take any values. When they are, we simply check that they equal themselves, which they will always do unless they are NA—and if they are, there isn't anything useful we can do about them anyway. Creating the test this way makes the code for testing patterns simpler, and testing that variables equal themselves does not change the result the conditions end up with.

We map this expression construction over all the patterns to get test-code for all recursive cases.

```
make_pattern_tests <- function(patterns, range_vars) {
    tests <- vector("list", length = length(patterns))
    for (i in seq_along(tests)) {
        tests[[i]] <- make_pattern_match(
            patterns[[i]],
            range_vars
        )
    }
    tests
}
```

Given these Fibonacci parser results:

```r
fib_ranges <- parse_ranges(rlang::quo({
    n <- 1:10
}))
fib_recursions <- parse_recursion(rlang::quo({
  F[n] <- F[n - 1] + F[n - 2] ? n > 2
  F[n] <- 1                   ? n <= 2
}))
```

we can construct the pattern tests for all cases like this:

```r
make_pattern_tests(
    fib_recursions$patterns,
    Map(as.symbol, names(fib_ranges))
)

## [[1]]
## all(n == n)
##
## [[2]]
## all(n == n)
```

We want to know if *both* pattern and ? condition match before we evaluate an expression, so we combine the two lists. We simply construct a new list where we combine the pattern matching with the ? conditions using &&.

```r
make_condition_checks <- function(
    ranges,
    patterns,
    conditions,
    recursions
) {
```

```
    test_conditions <- make_pattern_tests(
        patterns,
        Map(as.symbol, names(ranges))
    )
    for (i in seq_along(conditions)) {
        test_conditions[[i]] <- rlang::call2(
            "&&", test_conditions[[i]], conditions[[i]]
        )
    }
    test_conditions
}
```

To verify that this works as intended, we can, again, use the Fibonacci recursions.

```
fib_ranges <- parse_ranges(rlang::quo({
    n <- 1:10
}))
fib_recursions <- parse_recursion(rlang::quo({
  F[n] <- F[n - 1] + F[n - 2] ? n > 2
  F[n] <- 1                   ? n <= 2
}))
make_condition_checks(
    fib_ranges,
    fib_recursions$patterns,
    fib_recursions$conditions,
    fib_recursions$recursions
)

## [[1]]
## all(n == n) && n > 2
```

```
##
## [[2]]
## all(n == n) && n <= 2
```

So, the status is now that we have all the test conditions. We can take these test conditions, combine them with the expressions for evaluating the recursions, and construct a sequence of if-else expressions. If we construct an if call with two arguments, this is interpreted as the test condition and the value to compute when the test is true. If we construct a call with three arguments, the third is considered the else part of the statement. The simplest way to construct a sequence of if-else statements is, therefore, to start from the end of the list. We can translate the last case into an if statement without an else part—or we can make an else part that throws an error if we prefer. We can then iterate through the cases, at each case constructing a new if expression that takes the previous case as the else part. This idea can be implemented like this:

```
make_recursion_case <- function(
    test_expr,
    value_expr,
    continue
) {
    if (rlang::is_null(continue)) {
        rlang::call2("if", test_expr, value_expr)
    } else {
        rlang::call2("if", test_expr, value_expr, continue)
    }
}

make_update_expr <- function(
    ranges,
    patterns,
```

```
        conditions,
        recursions
) {
    conditions <- make_condition_checks(
        ranges,
        patterns,
        conditions,
        recursions
    )
    continue <- NULL
    for (i in rev(seq_along(conditions))) {
        continue <- make_recursion_case(
            conditions[[i]], recursions[[i]], continue
        )
    }
    continue
}
```

For the Fibonacci recursion, the result is this expression:

```
make_update_expr(
    fib_ranges,
    fib_recursions$patterns,
    fib_recursions$conditions,
    fib_recursions$recursions
)
## if (all(n == n) && n > 2) F[n - 1] + F[n - 2] else if (all(n ==
##     n) && n <= 2) 1
```

What remains to be done is to evaluate this expression for each combination of range indices. To do this, we create the table using the do. call expression we saw earlier, loop over all table rows, and insert the update expression inside the body of the loop. A function for creating the entire loop function looks like this[1]:

```
make_loop_expr <- function(tbl_name, update_expr) {
    rlang::expr({
        combs <- do.call(expand.grid, ranges)
        rlang::UQ(tbl_name) <- vector(
            "numeric",
            length = nrow(combs)
        )
        dim(rlang::UQ(tbl_name)) <- Map(length, ranges)
        for (row in seq_along(rlang::UQ(tbl_name))) {
            rlang::UQ(tbl_name)[row] <- with(
                combs[row, , drop = FALSE], {
                    rlang::UQ(update_expr)
                }
            )
        }
        rlang::UQ(tbl_name)
    })
}
```

In addition to the update_expr, the code needs to know the table name. This is because the update_expr will be referring to the table in the recursive cases. The table name is easy to get from the patterns, though.

---

[1]The drop=FALSE in the row subscript ensures that we get a row with a named variable even when there is only one.

```
get_table_name <- function(patterns) {
    p <- patterns[[1]]
    stopifnot(p[[1]] == "[")
    p[[2]]
}
get_table_name(fib_recursions$patterns)

## F
```

The loop, expanded with expressions from the Fibonacci recursion, looks like this:

```
make_loop_expr(get_table_name(fib_recursions$patterns),
               make_update_expr(
                    fib_ranges,
                    fib_recursions$patterns,
                    fib_recursions$conditions,
                    fib_recursions$recursions
               ))

## {
##     combs <- do.call(expand.grid, ranges)
##     F <- vector("numeric", length = nrow(combs))
##     dim(F) <- Map(length, ranges)
##     for (row in seq_along(F)) {
##         F[row] <- with(combs[row, , drop = FALSE], {
##             if (all(n == n) && n > 2)
##                 F[n - 1] + F[n - 2]
##             else if (all(n == n) && n <= 2)
##                 1
##         })
##     }
##     F
## }
```

For the edit-distance recursion, we get the following:

```
edit_ranges <- parse_ranges(rlang::quo({
    i <- 1:(length(x) + 1)
    j <- 1:(length(y) + 1)
}))
edit_recursions <- parse_recursion(rlang::quo({
  E[1,j] <- j - 1
  E[i,1] <- i - 1
  E[i,j] <- min(
      E[i - 1,j] + 1,
      E[i,j - 1] + 1,
      E[i - 1,j - 1] + (x[i - 1] != y[j - 1])
  )
}))

make_loop_expr(get_table_name(edit_recursions$patterns),
                  make_update_expr(
                      edit_ranges,
                      edit_recursions$patterns,
                      edit_recursions$conditions,
                      edit_recursions$recursions
                  ))

## {
##     combs <- do.call(expand.grid, ranges)
##     E <- vector("numeric", length = nrow(combs))
##     dim(E) <- Map(length, ranges)
##     for (row in seq_along(E)) {
##         E[row] <- with(combs[row, , drop = FALSE], {
##             if (all(1 == i, j == j) && TRUE)
##                 j - 1
```

```
##              else if (all(i == i, 1 == j) && TRUE)
##                 i - 1
##              else if (all(i == i, j == j) && TRUE)
##                 min(E[i - 1, j] + 1, E[i, j - 1] + 1, E[i - 1,
##                 j - 1] + (x[i - 1] != y[j - 1]))
##          })
##      }
##      E
## }
```

All that remains now is to evaluate this loop expression, and the only trick to this is to make sure we evaluate it in the right environment. The with statement inside the loop ensures that we over-scope the expression with the relevant index variables, so we do not have to worry about the ranges and range indices. The expressions we evaluate in the update expression, however, should be evaluated in the scope where we define the expression, which we captured in the quosures when we parsed the expressions. Ideally, we would just evaluate the loop in this quosure environment, but the loop expression needs to know about the ranges, so we need to put those in the scope as well. One way to do this is to put a local function-call environment between the quosure scope and the loop and evaluate the expression this way:

```
eval_recursion <- function(ranges, recursions) {
    loop <- make_loop_expr(
                get_table_name(recursions$patterns),
                make_update_expr(
                    ranges,
                    recursions$patterns,
                    recursions$conditions,
                    recursions$recursions
                ))
```

```
    eval_env <- rlang::env_clone(
        environment(), # this function environment
        recursions$recursion_env # quosure environment
    )
    eval(loop, envir = eval_env)
}
```

For the Fibonacci expressions, we get the following:

```
eval_recursion(fib_ranges, fib_recursions)
```

```
## [1]  1  1  2  3  5  8 13 21 34 55
```

The %where% operator, as we saw earlier in this chapter, simply parses the ranges and recursions and then evaluates the result in a function called eval_dynprog. This function does nothing more than call the eval_recursion function we just wrote.

```
eval_dynprog <- function(dynprog) {
    eval_recursion(dynprog$ranges, dynprog$recursions)
}
```

Providing the parsed pieces directly to the eval_dynprog function, instead of going through %where%, we see that this is indeed what is happening.

```
eval_dynprog(list(
    ranges = fib_ranges,
    recursions = fib_recursions
))
```

```
## [1]  1  1  2  3  5  8 13 21 34 55
```

```
eval_dynprog(list(
    ranges = edit_ranges,
    recursions = edit_recursions
))
```

```
##        [,1] [,2] [,3] [,4] [,5]
## [1,]     0    1    2    3    4
## [2,]     1    0    1    2    3
## [3,]     2    1    0    1    2
## [4,]     3    2    1    1    1
```

Fixing the Evaluation Environment

The solution so far works, but there is one slightly unsatisfactory aspect of the solution. When we evaluate the loop, we place an environment between the quosure environment and the expression we evaluate, and this environment contains several local variables that can potentially conflict with variables in the quosure scope.

We can avoid this and make sure that we only over-scope with the range-index variables and nothing else, but we will then have to take a slightly different approach. Instead of constructing the loop as an expression and then evaluating it in the function-plus quosure-scope, we loop over the range-index combinations immediately. In the body of the loop, we construct and evaluate the update expressions.[2]

The alternative solution is listed next. As in the previous solution, we get the name of the dynamic-programming table first, and we construct the update expression. This time, though, we also get hold of the table name as a string. We need this to access the name in the evaluation environment, which we construct as an empty environment with the quosure environment as the parent. That way, we can modify the evaluation environment without affecting the quosure environment, but we still have access to all variables we can get from there.

---

[2]Depending on your taste, you might prefer this solution over the previous just because we keep evaluation of generated code to a minimum. The new solution involves generating and evaluating more expressions, though, and some hacks to update the table in the scope where we evaluate recursions, so the first solution has some merits over the second.

We don't need any meta-programming to construct the table. We can get its size and dimensions from the ranges. We need to put it in the evaluation environment for the update expressions to see, though, and we do that using the [[operator on environments. After constructing the table and putting it in the evaluation environment, we loop through all the range-index combinations as before.

We want to over-scope the update expression with the index variables when we compute the recursion values, so we need to evaluate the expression in an environment that contains combs[row,]. We cannot evaluate the actual assignment if we use this row as the second argument to eval, however. R considers the second argument to eval the environment where expressions are evaluated, and an assignment in an environment that is really a list will not work. You will not get an error, but you will not assign to any variable either. So, we need to split the evaluation of the recursion and the assignment to the table in two. We evaluate the recursion expression in an environment where combs[row,] over-scopes the evaluation environment, and we evaluate the assignment in the evaluation environment without any other over-scoping.

You might think that once we have the value to insert in the table, we could just update the table, and we wouldn't have to construct an assignment expression to achieve this. It is not that simple, however. R tries very hard to make data immutable, and if we assign to an element in a table that is referenced from two different places, R will copy the table, update only one of the copies, and set the modified copy in the environment where we did the assignment, leaving the original in the other scope. To update the table, such that the recursion expressions can use it, we need to do the update in the evaluation environment (or alternatively, update the table and then assign the copy into the evaluation environment).

The alternative evaluation function thus looks like this:

```
eval_recursion <- function(ranges, recursions) {
    tbl_name <- get_table_name(recursions$patterns)
    tbl_name_string <- as.character(tbl_name)
    update_expr <- make_update_expr(
        ranges,
        recursions$patterns,
        recursions$conditions,
        recursions$recursions
    )
    eval_env <- rlang::child_env(recursions$recursion_env)

    combs <- do.call(expand.grid, ranges)
    tbl <- vector("numeric", length = nrow(combs))
    dim(tbl) <- Map(length, ranges)
    eval_env[[tbl_name_string]] <- tbl

    for (row in seq_along(tbl)) {
        val <- eval(
            rlang::expr(rlang::UQ(update_expr)),
            combs[row, , drop = FALSE],
            eval_env
        )
        eval(rlang::expr(
                rlang::UQ(tbl_name)[rlang::UQ(row)]
                    <- rlang::UQ(val)
            ), eval_env)
    }

    eval_env[[tbl_name_string]]
}
```

It behaves exactly like the previous one, though.

```
eval_dynprog(list(
    ranges = fib_ranges,
    recursions = fib_recursions
))
```

```
## [1]  1  1  2  3  5  8 13 21 34 55
```

```
eval_dynprog(list(
    ranges = edit_ranges,
    recursions = edit_recursions
))
```

```
##       [,1] [,2] [,3] [,4] [,5]
## [1,]    0    1    2    3    4
## [2,]    1    0    1    2    3
## [3,]    2    1    0    1    2
## [4,]    3    2    1    1    1
```

At least, unless your recursions referenced variables that would be over-scoped by local variables in the previous version.

.

CHAPTER 13

Conclusion

Embedding domain-specific languages into R enriches the language by providing flexible ways to construct data structures or process data. In this book, we have seen how to use techniques such as operator overloading, meta-programming, and non-standard evaluation to implement small domain-specific languages. Through examples, we have explored several language designs for various problems.

The flexibility of the R language—flexibility both in how we can override operators and functions and specialize generic functions and flexibility in how we evaluate expressions—makes domain-specific languages a natural approach to designing package interfaces. The packages in the widely popular tidyverse exploit this to a great degree. The ggplot2 package uses the plus operator to combine graphical commands in the "grammar of graphics." The magrittr pipe operator, %>%, is used to create sequences of transformation verbs in packages such as dplyr and tidyr. The tidyverse packages also exploit non-standard evaluation to over-scope expressions with data frame columns. As is evident from the popularity of the tidyverse, the use of well-designed domain-specific languages can improve the productivity of a programmer considerably. This happens as a consequence of improved readability and increased flexibility over more traditional interfaces to analysis frameworks.

© Thomas Mailund 2018
T. Mailund, *Domain-Specific Languages in R*,
https://doi.org/10.1007/978-1-4842-3588-1_13

A key word here is *well-designed*. When we implement a domain-specific language, rather than rely on standard R syntax, we demand of the user that he or she can use the language more efficiently than they could use a more traditional interface. This will be the case only if the language is designed to be consistent in how you combine its components and how it interacts with the surrounding R code.

Designing a language requires some trial and error. With experience, you will reduce the number of errors, of course, but you will always benefit from experimenting with alternative ways of expressing the same ideas. It is via experiments you will learn how different components of a language combine to express ideas. A good approach to designing a new domain-specific language is to consider various use cases and write down how you would ideally want to express the computations you want to implement in your new language. You don't have to implement any of the language constructs yet; you can just try to write down expressions in various ways.

Once you are satisfied with how you would ideally want the language to look, you can start to worry about how you would implement it. It might not be possible to implement the language to accept exactly the syntax you came up with in the design phase. Since you are embedding the language in R, expressions in the language must also be valid R code. This puts some restrictions on what you can do. But starting from the ideal design, you can modify your use cases until the examples are valid expressions both in your language and in R, and from there, you can exploit the techniques you have learned in this book to implement your language.

Good luck.

References

Germane, K, and M Might. 2014. "Deletion: The Curse of the Red-Black Tree." *Journal of Functional Programming* 24 (04): 423–33.

Mailund, T. 2017a. *Functional Data Structures in R: Advanced Statistical Programming in R.* Apress.

———. 2017b. *Functional Programming in R: Advanced Statistical Programming for Data Science, Analysis and Finance.* Apress.

———. 2017c. *Metaprogramming in R: Advanced Statistical Programming for Data Science, Analysis and Finance.* Apress.

Okasaki, C. 1999. "Red-Black Trees in a Functional Setting." *Journal of Functional Programming* 9 (4). Cambridge University Press: 471–77.

© Thomas Mailund 2018
T. Mailund, *Domain-Specific Languages in R*,
https://doi.org/10.1007/978-1-4842-3588-1

Index

A, B

Abstract syntax tree (AST), 34–35
Assignment operator, 104

C

Constructors
    as_list(environment()), 191
    binary tree, 194
    bind_rows function, 190
    construction_printer function,
        189, 193
    constructor_constant
        function, 189
    deparse_construction and
        construction_printer
        functions, 187
    DSL, 186
    environment, 192
    is_lang function, 189
    pretty-printing elements, 186
    process_alternatives function,
        187–188
    process_arg function, 190
    process_constructor_function,
        188–189, 191
    string representation, 193

toString function, 193
    variables and function, 185
Continuous-time Markov chains
    (CTMCs)
    add_edge function, 170–171
    class creation, 170
    collect_symbols function, 169
    collect_symbols_rec and
        make_args_list, 167, 169
    expm package, 167
    foo to qux edges, 173
    functions, 49
    likelihood function, 180, 182
    magrittr package, 167
    mathematical modeling, 48
    parameters, 171–172
    probability vectors flow, 52
    rate matrix, 48–50, 53,
        174–175, 177
    tibble package, 167
    trace, 177–178, 180
    transition probabilities, 51

D

Default environment, 126
Default parameters

© Thomas Mailund 2018
T. Mailund, *Domain-Specific Languages in R*,
https://doi.org/10.1007/978-1-4842-3588-1

Q, R

Printed in the United States
By Bookmasters